8,202
798
LC Record

MAY 26 '11

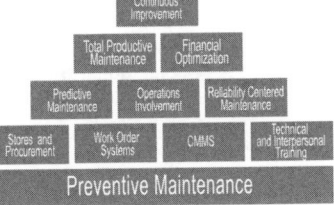

Training Programs for Maintenance Organizations

Terry Wireman, CPMM

www.terrywireman.com
TLWireman@Mindspring.com
twireman@vestapartners.com

Industrial Press, Inc.

Library of Congress Cataloging-in-Publication Data

Wireman, Terry.
  Training programs for maintenance organizations / Terry Wireman.
    p. cm. -- (Maintenance strategy series ; v. 5)
  Includes bibliographical references and index.
  ISBN 978-0-8311-3369-6 (hardcover)
  1. Industrial equipment--Maintenance and repair--Study and teaching. 2.
Plant maintenance--Study and teaching. 3. Occupational training. I. Title.

TS192.W576 2009
658.2'02--dc22

                                           2009013417

Industrial Press, Inc.
989 Avenue of the Americas
New York, NY 10018

Sponsoring Editor: John Carleo
Interior Text and Cover Design: Janet Romano
Developmental Editor: Robert Weinstein

10   9   8   7   6   5   4   3   2   1

TABLE OF CONTENTS

iii

iv

INTRODUCTION

Volume 5 Training Programs for Maintenance Organizations

Volume 5 of the Maintenance Strategy Series builds on the previous four volumes. One might wonder why we have waited so long to address training for maintenance organizations; however, many of the strategies discussed in the previous four volumes must be in place to provide the information necessary to developing good training programs. As will be shown in this volume, the strategies discussed previously build the needs foundation for the training requirements.

It becomes critical at this point in the development of maintenance strategies that the skills of the individuals in the organization be addressed. It will be virtually impossible to move a maintenance organization into the predictive and reliability-based programs without guaranteeing a solid skill foundation. In addition, much of the training that is to be developed can be utilized to train operators, who will be involved in performing some basic maintenance tasks. The additional utilization of the training will be highlighted in Volume 6 of the Maintenance Strategy Series, *Operations and Team-Based Maintenance Activities*, which also highlights team-based maintenance activities.

In Volume 5, we will discuss the current problems facing maintenance workforces. Rather than presenting these problems as insurmountable obstacles, the text will develop the process to overcome them. It begins with a foundation training program that should be required for all skilled trades personnel. In addition, the text will look at how to develop advanced training that addresses specific trade needs of the trainees. Further, the text will consider how to develop equipment specific training that will enable companies to fully utilize their assets.

Although many of the concepts have been in the training realm for years, they have not been directly applied to training a technical workforce. This text will make that application. In addition to the theoretical material, the text will present actual examples from existing technical training programs. Many of the examples related in this volume will be based on the experience the author has gained developing and implementing a comprehensive maintenance skills training program while working at a major integrated steel company. This program processed approximately 200 individuals from apprentices to journeyman level technicians every four years. Additional examples will be drawn from client projects where similar comprehensive technical skills training programs have been developed.

OVERVIEW

The Maintenance Strategy Series Process Flow

Good, sound, functional maintenance practices are essential for effective maintenance / asset management strategies. But what exactly are "good, sound, functional maintenance practices?" The materials contained in this overview (and the overview for each of the volumes in the Maintenance Strategy Series) explain each block of the Maintenance Strategy Series Process Flow. They are designed to highlight the steps necessary to develop a complete maintenance / asset management strategy for your plant or facility. The activities described in the Process Flow are designed to serve as a guide for strategic planning discussions. The flow diagram for the Maintenance Strategy Series Process Flow can be found at the end of this overview.

Author's Note

Many individuals may believe that this type of maintenance strategy program is too expensive or time consuming to implement, especially when there are advanced predictive or reliability techniques that might be employed. Yet there is a reason for the sequencing of the Maintenance Strategy Series process flow. If attempts are made to deploy advanced techniques before the organization is mature enough to properly understand and utilize them (basically, the "I want results now" short-term focus), they will fail. The reason? Developing and implementing a sustainable maintenance / asset management strategy is more than just distributing a flow chart or dictionary of technical terms. It is an educational exercise that must change a company culture. The educational process that occurs during a structured implementation of basic maintenance processes must evolve into more sophisticated and advanced processes as the organization develops the understanding and skills necessary.

If an individual is to obtain a college degree, it may involve an investment of four or more years to achieve this goal. Likewise, if a company is to obtain an advanced standing in a maintenance / asset management strategy, it may take up to four years. It is not that someone cannot,

3

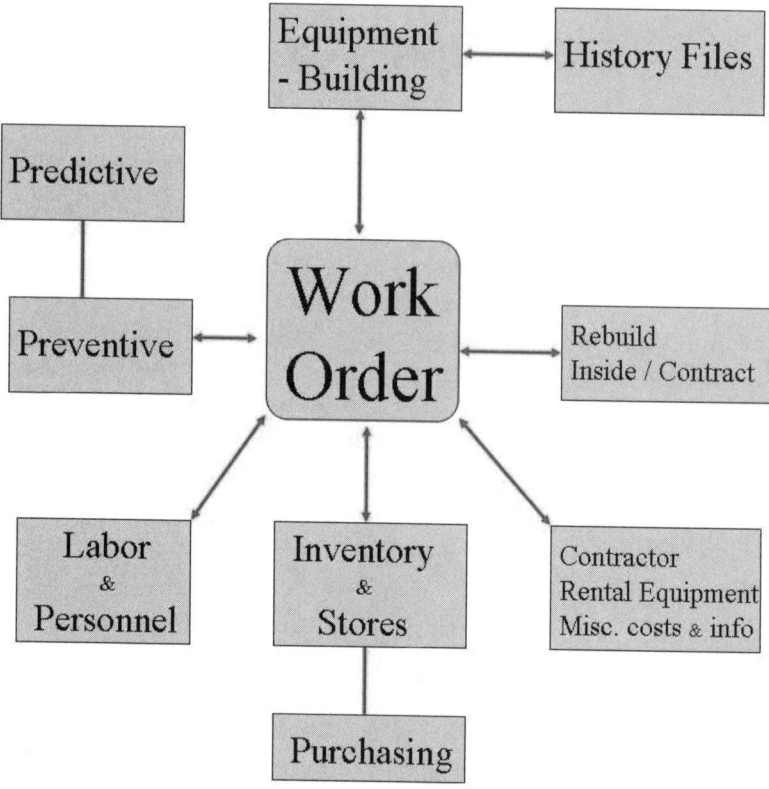

Figure I-4 Work Order and MRO Cost Relationships

through years of experience and education, design their maintenance / asset management process in a short time period. It will, however, take the entire organization (from senior executive to shop floor employees) this amount of time to become mature in their understanding and utilization of the process. Although there will be incremental benefits achieved along the journey to maintenance / asset management excellence, the true benefits are not realized until there is a complete organizational focus on maximizing all aspects of the investment in the assets. It is this competitive focus that separates long-term, sustainable success from a short-term "flash" of improvement.

In the beginning, it is necessary for a plant or facility to decide it is necessary to improve their maintenance / asset management strategy. The business reason for the needed improvement can be multi-faceted, but would likely include:

- Poor Return On Investment (ROI) for the total plant or facility valuation
- Poor throughput for the design of the plant
- Inability to meet production demands
- High cost of occupancy for a facility
- Excessive downtime
- Production inefficiencies

Once the decision has been made to develop / improve the maintenance/ asset management strategy, the Maintenance Strategy Series process flow diagram should be followed. It begins with Preventive Maintenance.

1. Does a PM Program Exist?
Preventive maintenance is the core of any equipment/asset maintenance process improvement strategy. All plant and facility equipment, including special back-up or redundant equipment, must be covered by a complete, cost-effective, preventive maintenance program. The preventive maintenance program must be designed to eliminate all unplanned equipment failures. The preventive maintenance program should be designed to insure proper coverage of the critical equipment of the plant or facility. The

MRO Inventory Best Practices

- 95 - 97% Service Levels
- 100% accuracy of data
- > 1 turn per year on inventory value
- Elimination of non-moving spares
- Reduction of slow moving spares
- Controlled access
- Consignment arrangements
- Strategic partnering with suppliers

Figure I-5 MRO Inventory Best Practices

program should include a good cross section of the following:
- Inspections
- Adjustments
- Lubrication
- Proactive replacements of worn components

The goal of the program is to insure there will be no unplanned equipment downtime.

2. Is the PM Program Effective?

The effectiveness of the preventive maintenance program is determined by the level of unplanned equipment maintenance that is performed. Unplanned equipment maintenance is defined as any maintenance activity that is performed with less than one week of advanced planning. Unplanned equipment maintenance is commonly referred to as reactive maintenance. An effective preventive maintenance program will reduce the amount of unplanned work to less than 20% of the total manpower expended for all equipment maintenance activities. If more time than this is being spent on unplanned activities, then a reevaluation of the preventive maintenance program is required. It will take more resources and additional time to make progress in any of the following maintenance process areas unless the preventive maintenance program is effective enough for the equipment maintenance to meet the 80%/20% rule.

It should be the goal of not progressing any further until the preventive maintenance program is successful. In addition to requiring more resources and taking longer to develop the subsequent maintenance processes, it is very common to see companies try to compensate for a reactive organization. This means they will circumvent some of the "best practices' in the subsequent processes to make them work in a reactive environment. All this will do is reinforce negative behavior and sub-optimize the effectiveness of the subsequent processes.

3. Do MRO Processes Exist?

After the preventive maintenance program is effective, the equipment spares, stores, and purchasing systems must be analyzed. The equipment spares and stores should be organized, with all of the spares identified and tagged, stored in an identified location, with accurate on-hand and usage data. The purchasing system must allow for procurement of all necessary spare parts to meet the maintenance schedules. All data necessary to track the cost and usage of all spare parts must be complete and accurate.

4. Are the MRO Processes Effective?

The benchmark for an effective maintenance / asset management MRO process is service level. Simply defined, the service level measures what percent of the time a part is in stock when it is needed. The spare parts must be on hand at least 95%– 97% of the time for the stores and purchasing systems to support the maintenance planning and scheduling functions.

Again, unless maintenance activities are proactive (less than 20% unplanned weekly), it will be impossible for the stores and purchasing groups to be cost effective in meeting equipment maintenance spare parts demands. They will either fall below the 95%–97% service level, or they will be forced to carry excess inventory to meet the desired service level.

The MRO process must be effective for the next steps in the strategy development. If the MRO data required to support the maintenance work management process is not developed, the maintenance spare parts costs will never be accurate to an equipment level. The need for this level of data accuracy will be explained in Sections 6 and 10 of the preface.

5. Does a Work Management Process Exist?

The work management system is designed to track all equipment maintenance activities. The activities can be anything from inspections and adjustments to major overhauls. Any maintenance that is performed without being recorded in the work order system is lost. Lost or unrecorded data makes it impossible to perform any analysis of equipment problems. All activities performed on equipment must be recorded to a work order by the responsible individual. This highlights the point that maintenance, operations, and engineering will be extremely involved in utilizing work orders.

Beyond just having a work order, the process of using a work order system needs detailed. A comprehensive work management process should include details on the following:

- How to request work
- How to prioritize work
- How to plan work
- How to schedule work
- How to execute work

- How to record work details
- How to process follow up work
- How to analyze historical work details

6. Is the Work Management Process Effective?

This question should be answered by performing an evaluation of the equipment maintenance data. The evaluation may be as simple as answering the following questions:

- How complete is the data?
- How accurate is the data?
- How timely is the data?
- How usable is the data?

If the data is not complete, it will be impossible to perform any meaningful analysis of the equipment historical and current condition. If the data is not accurate, it will be impossible to correctly identify the root cause of any equipment problems. If the data is not timely, then it will be impossible to correct equipment problems before they cause equipment failures. If the data is not usable, it will be impossible to format it in a manner that allows for any meaningful analysis. Unless the work order system provides data that passes this evaluation, it is impossible to make further progress.

7. Is Planning and Scheduling Utilized?

This review examines the policies and practices for equipment maintenance planning and scheduling. Although this is a subset of the work management process, it needs a separate evaluation. The goal of planning and scheduling is to optimize any resources expended on equipment maintenance activities, while minimizing the interruption the activities have on the production schedule. A common term used in many organizations is "wrench time." This refers to the time the craft technicians have their hands on tools and are actually performing work; as opposed to being delayed or waiting to work. The average reactive organization may have a wrench time of only 20%, whereas a proactive, planned, and scheduled organization may be as high as 60% or even a little more.

The ultimate goal of planning and scheduling is to insure that all equipment maintenance activities occur like a pit stop in a NASCAR race. This insures optimum equipment uptime, with quality equipment maintenance activities being performed. Planning and scheduling pulls together all of the activities, (maintenance, operations, and engineering) and focuses them on obtaining maximum (quality) results in a minimum amount of time.

8. Is Planning and Scheduling Effective?

Although this question is similar to #6, the focus is on the efficiency and effectiveness of the activities performed in the 80% planned mode. An efficient planning and scheduling program will insure maximum productivity of the employees performing any equipment maintenance activities. Delays, such as waiting on or looking for parts, waiting on or looking for rental equipment, waiting on or looking for the equipment to be shut down, waiting on or looking for drawings, waiting on or looking for tools, will all be eliminated.

If these delays are not eliminated through planning and scheduling, then it will be impossible to optimize equipment utilization. It will be the same as a NASCAR pit crew taking too long to do a pit stop; the race is lost by not keeping the car on the track. The equipment utilization is lost by not properly keeping the equipment in service.

9. Is a CMMS / EAM System Utilized?

By this point in the Maintenance Strategy Process development, a considerable volume of data is being generated and tracked. Ultimately, the data becomes difficult to manage using manual methods. It may be necessary to computerize the work order system. If the workforce is burdened with excessive paper work and is accumulating file cabinets of equipment data that no one has time to look at, it is best to computerize the maintenance / asset management system. The systems that are used for managing the maintenance /asset management process are commonly referred to by acronyms such as CMMS (Computerized Maintenance Management Systems) or EAM (Enterprise Asset Management) systems. (The difference between the two types of systems will be thoroughly covered in Volume Four.)

The CMMS/ EAM System should be meeting the equipment management information requirements of the organization. Some of the requirements include:
- Complete tracking of all repairs and service
- The ability to develop reports, for example:
- Top ten equipment problems

- Most costly equipment to maintain
- Percent reactive vs. proactive maintenance
- Cost tracking of all parts and costs

If the CMMS/EAM system does not produce this level of data, then it needs to be re-evaluated and a new one may need to be implemented.

10. Is the CMMS/ EAM System Utilization Effective?

The re-evaluation of the CMMS / EAM system may also highlight areas of weakness in the utilization of the system. This should allow for the specification of new work management process steps that will correct the problems and allow for good equipment data to be collected. Several questions for consideration include:

- Is the data we are collecting complete and accurate?
- Is the data collection effort burdening the work force?
- Do we need to change the methods we use to manage the data?

Once problems are corrected and the CMMS / EAM system is being properly utilized, then constant monitoring for problems and solutions must be put into effect.

The CMMS / EAM system is a computerized version of a manual system. There are currently over 200 commercially produced CMMS / EAM systems in the North American market. Finding the correct one may take some time, but through the use of lists, surveys, and "word of mouth," it should take no more than three to a maximum of six months for any organization to select their CMMS / EAM system. When the right CMMS / EAM system is selected, it then must be implemented. CMMS / EAM system implementation may take from three months (smaller organizations) to as long as 18 months (large organizations) to implement. Companies can spend much time and energy around the issue of CMMS selection and implementation. It must be remembered that the CMMS / EAM system is only a tool to be used in the improvement process; it is not the goal of the process. Losing sight of this fact can curtail the effectiveness of any organization's path to continuous improvement.

If the correct CMMS / EAM system is being utilized, then it makes the equipment data collection faster and easier. It should also make the analysis of the data faster and easier. The CMMS / EAM system should assist in enforcing "World Class" maintenance disciplines, such as planning and scheduling and effective stores controls. The CMMS / EAM system should provide the employees with usable data with which to make equipment management decisions. If the CMMS / EAM system is not improving these efforts, then the effective usage of the CMMS / EAM system needs to be evaluated. Some of the problems encountered with CMMS / EAM systems include:

- Failure to fully implement the CMMS
- Incomplete utilization of the CMMS
- Inaccurate data input into the CMMS
- Failure to use the data once it is in the CMMS

11. Do Maintenance Skills Training Programs Exist?

This question examines the maintenance skills training initiatives in the company. This is a critical item for any future steps because the maintenance organization is typically charged with providing training for any operations personnel that will be involved in future activities. Companies need to have an ongoing maintenance skills training program because technology changes quickly. With newer equipment (or even components) coming into plants almost daily, the skills of a maintenance workforce can be quickly dated. Some sources estimate that up to 80% of existing maintenance skills can be outdated within five years. The skills training program can utilize many resources, such as vocational schools, community colleges, or even vendor training. However, to be effective, the skills training program needs to focus on the needs of individual employees, and their needs should be tracked and validated.

12. Are the Maintenance Training Programs Effective?

This evaluation point focuses on the results of the skills training program. It deals with issues such as:

- Is there maintenance rework due to the technicians not having the skills necessary to perform the work correctly the first time?
- Is there ongoing evaluation of the employees skills versus the new technology or new equipment they are being asked to main tain or improve?
- Is there work being held back from certain employees because a manager or supervisor questions their ability to complete the work in a timely or quality manner?

If these questions uncover some weaknesses in the workforce, then it quickly shows that the maintenance skills training program is not effective. If this is the case, then a duty-task-needs analysis will highlight the content weaknesses in the current maintenance skills training program and provide areas for improvement to increase the versatility and utilization of the maintenance technicians.

13. Are Operators Involved in Maintenance Activities?

As the organization continues to make progress in the maintenance disciplines, it is time to investigate whether operator involvement is possible in some of the equipment management activities. There are many issues that need to be explored, from the types of equipment being operated, the operators-to-equipment ratios, and the skill levels of the operators, to contractual issues with the employees' union. In most cases, some level of activity is found in which the operators can be involved within their areas. If there are no obvious activities for operator involvement, then a re-evaluation of the activities will be necessary.

The activities the operators may be involved in may be basic or complex. It is partially determined by their current operational job requirements. Some of the more common tasks for operators to be involved in include, but are not necessarily limited to:

    a. Equipment Cleaning: This may be simply wiping off their equipment when starting it up or shutting it down.

    b. Equipment Inspecting: This may range from a visual inspection while wiping down their equipment to a maintenance inspections checklist utilized while making operational checks.

    c. Initiating Work Requests: Operators may make out work requests for any problems (either current or developing) on their equipment. They would then pass these requests on to maintenance for entry into the work order system. Some operators will directly input work requests into a CMMS.

    d. Equipment Servicing: This may range from simple running adjustments to lubrication of the equipment.

    e. Visual Systems: Operators may use visual control techniques to inspect and to make it easier to determine the condition of their equipment.

Whatever the level of operator involvement, it should contribute to the improvement of the equipment effectiveness.

14. Are Operator-Performed Maintenance Tasks Effective?

Once the activities the operators are to be involved with has been determined, their skills to perform these activities need to be examined. The operators should be properly trained to perform any assigned tasks. The training should be developed in a written and visual format. Copies of the training materials should be used when the operators are trained and a copy of the materials given to the operators for their future reference. This will contribute to the commonality required for operators to be effective while performing these tasks. It should also be noted that certain regulatory organizations require documented and certified training for all employees (Lock Out Tag Out is an example).

Once the operators are trained and certified, they can begin performing their newly-assigned tasks. It is important for the operators to be coached for a short time to insure they have the full understanding of the hows and whys of the new tasks. Some companies have made this coaching effective by having the maintenance personnel assist with it. This allows for operators to receive background knowledge that they may not have gotten during the training.

15. Are Predictive Techniques Utilized?

Once the operators have begun performing some of their new tasks, some maintenance resources should be available for other activities. One area that should be explored is predictive maintenance. Some fundamental predictive maintenance techniques include:
- Vibration Analysis
- Oil Analysis
- Thermography
- Sonics

Plant equipment should be examined to see if any of these techniques will help reduce downtime and improve its service. Predictive technologies should not be utilized because they are technically advanced, but only when they contribute to improving the equipment effectiveness. The correct technology should be used to trend or solve the equipment problems encountered.

16. Are the PDM Tasks Effective?

If the proper PDM tools and techniques are used, there should be a decrease in the downtime of the equipment. Because the PDM program will find equipment wear before the manual PM techniques, the planning and scheduling of maintenance activities should also increase. In addition, some of the PM tasks that are currently being performed at the wrong interval should also be able to be adjusted. This will have a positive impact on the cost of the PM program. The increased efficiency of the maintenance workforce and the equipment should allow additional time to focus on advanced reliability techniques.

17. Are Reliability Techniques Being Utilized?

Reliability Engineering is a broad term that includes many engineering tools and techniques. Some common tools and techniques include:

a. **Life Cycle Costing:** This technique allows companies to know the cost of their equipment from when it was designed to the time of disposal.

b. **R.C.M.:** Reliability Centered Maintenance is used to track the types of maintenance activities performed equipment to insure they are correct activities to be performed.

c. **F.E.M.A.:** Failure and Effects Mode Analysis examines the way the equipment is operated and any failures incurred during the operation to find methods of eliminating or reducing the numbers of failures in the future.

d. **Early Equipment Management and Design:** This technique takes information on equipment and feeds it back into the design process to insure any new equipment is designed for maintain ability and operability.

Using these and other reliability engineering techniques improve equipment performance and reliability to insure competitiveness.

18. Are the Reliability Techniques Effective?

The proper utilization of reliability techniques will focus on eliminating repetitive failures on the equipment. While some reliability programs will also increase the efficiency of the equipment, this is usually the focus of TPM/OEE techniques. The elimination of the repetitive failures will increase the availability of the equipment. The effectiveness of the reliability techniques are measured by maximizing the uptime of the equipment.

19. Are TPM/ OEE Methodologies Being Utilized?

Are the TPM/OEE methodologies being utilized throughout the company? If they are not, then the TPM/OEE program needs to be examined for application in the company's overall strategy. If a TPM/OEE process exists, then it should be evaluated for gaps in performance or deficiencies in existing parts of the process. Once weaknesses are found, then steps should be taken to correct or improve these areas. Once the weaknesses are corrected and the goals are being achieved, then the utilization of the OEE for all equipment relate decisions is examined.

20. Is OEE Being Effectively Utilized?

The Overall Equipment Effectiveness provides a holistic look at how the equipment is utilized. If the OEE is too low, it indicates that the equipment is not performing properly and maximizing the return on investment in the equipment. Also, the upper limit for the OEE also needs to be understood. If a company were to focus on achieving the maximum OEE number, they may pay too much to ever recover the investment. If the OEE is not clearly understood, then additional training in this area must be provided. Once the OEE is clearly understood, then the focus can be switched to achieving the financial balance required to maximize a company's return on assets (ROA).

21. Does Total Cost Management Exist?

Once the equipment is correctly engineered, the next step is to understand how the equipment or process impacts the financial aspects of the company's business. Financial optimization considers all costs impacted when equipment decisions are made. For example, when calculating the timing to perform a preventive maintenance task, is the cost of lost production or downtime considered? Are wasted energy costs considered when cleaning heat exchangers or coolers? In this step, the equipment data collected by the company is examined in the context of the financial impact it has on the company's profitability. If the data exists and the information systems are in place to continue to collect the data, then financial optimization should be utilized. With this tool, equipment teams will be able to financially manage their equipment and processes.

22. Is Total Cost Management Utilized?

While financial optimization is not a new technique, most companies do not properly utilize it because they do not have the data necessary to make the technique effective. Some of the data required includes:

- MTBF (Mean Time Between Failure) for the equipment
- MTTR (Mean Time To Repair) for the equipment
- Downtime or lost production costs per hour
- A Pareto of the failure causes for the equipment
- Initial cost of the equipment
- Replacement costs for the equipment
- A complete and accurate work order history for the equipment

Without this data, financial optimization can not be properly conducted on equipment. Without the information systems in place to collect this data, a company will never have the accurate data necessary to perform financial optimization.

23. Are Continuous Improvement Techniques Utilized for Maintenance / Asset Management Decisions?

Once a certain level of proficiency is achieved in maintenance/ asset management, companies can begin to lose focus on their improvement efforts. They may even become complacent in their improvement efforts. However, there are excellent Continuous Improvement (CI) tools for examining even small problems. If new tools are constantly examined and applied to existing processes, all opportunities for improvement will be clearly identified and prioritized.

24. Are the CI Tools and Techniques Effective?

This question may appear to be subjective; however, improvements at this phase of maturity for a maintenance / reliability effort may be small and difficult to identify. However, the organizational culture of always looking for areas to improve is a true measure of the effectiveness of this step. As long as even small improvements in maintenance / reliability management are realized, this question should be answered "Yes."

25. Is Continuous Improvement Sought After in All Aspects of Maintenance / Asset Management?

When organizations reaches this stage, it will be clear that they are leaders in maintenance / reliability practices. Now, they will need continual focus on small areas of improvement. Continuous improvement means never getting complacent. It is the constant self-examination with the focus on how to become the best in the world at the company's business. Remember:

Yesterday's Excellence

is

Today's Standard

and

Tomorrow's Mediocrity

Maintenance Strategy Series Part 1

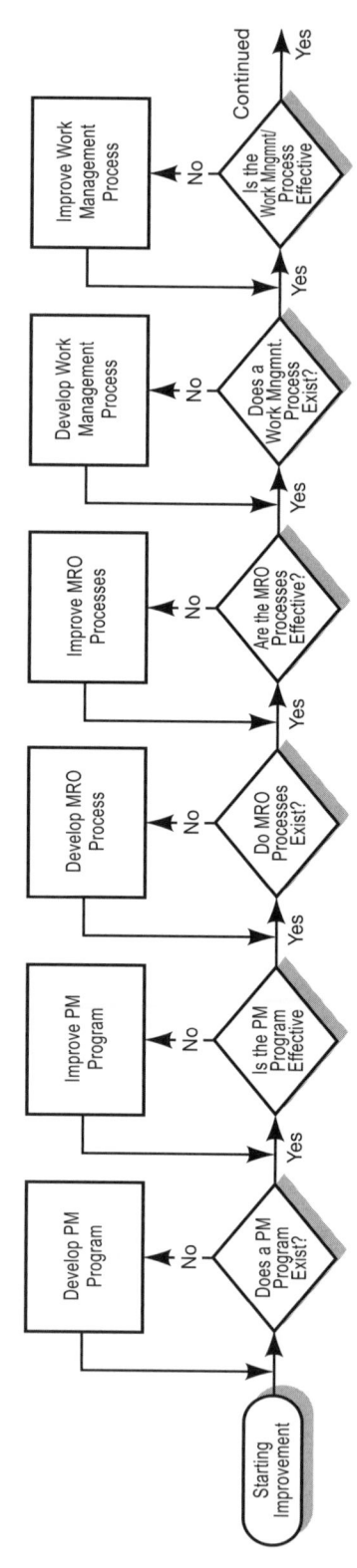

Maintenance Strategy Series Part 2

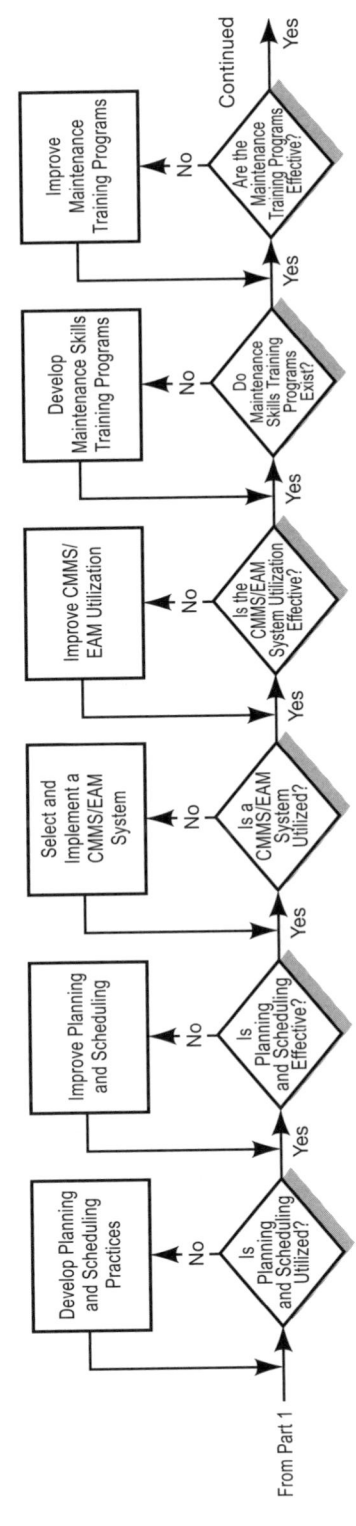

Maintenance Strategy Series Part 3

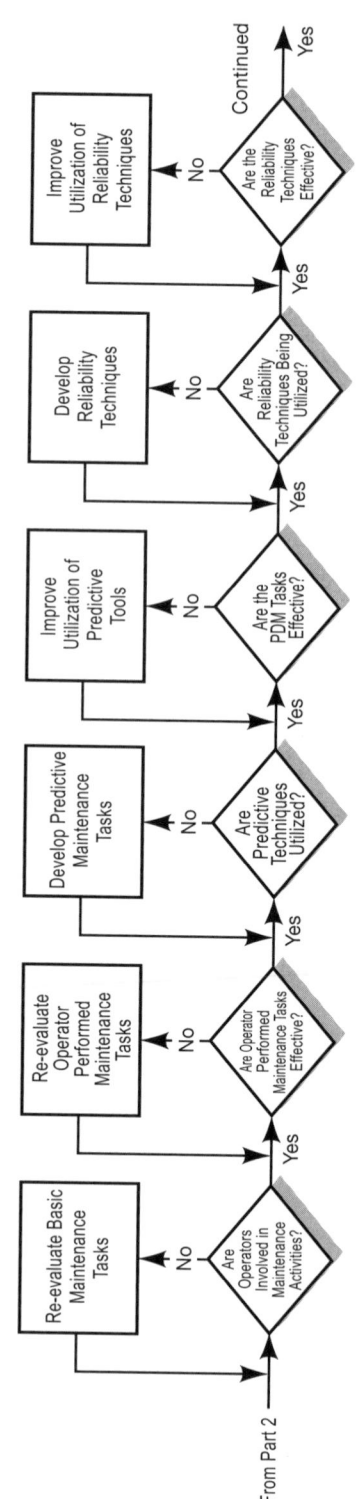

Maintenance Strategy Series Part 4

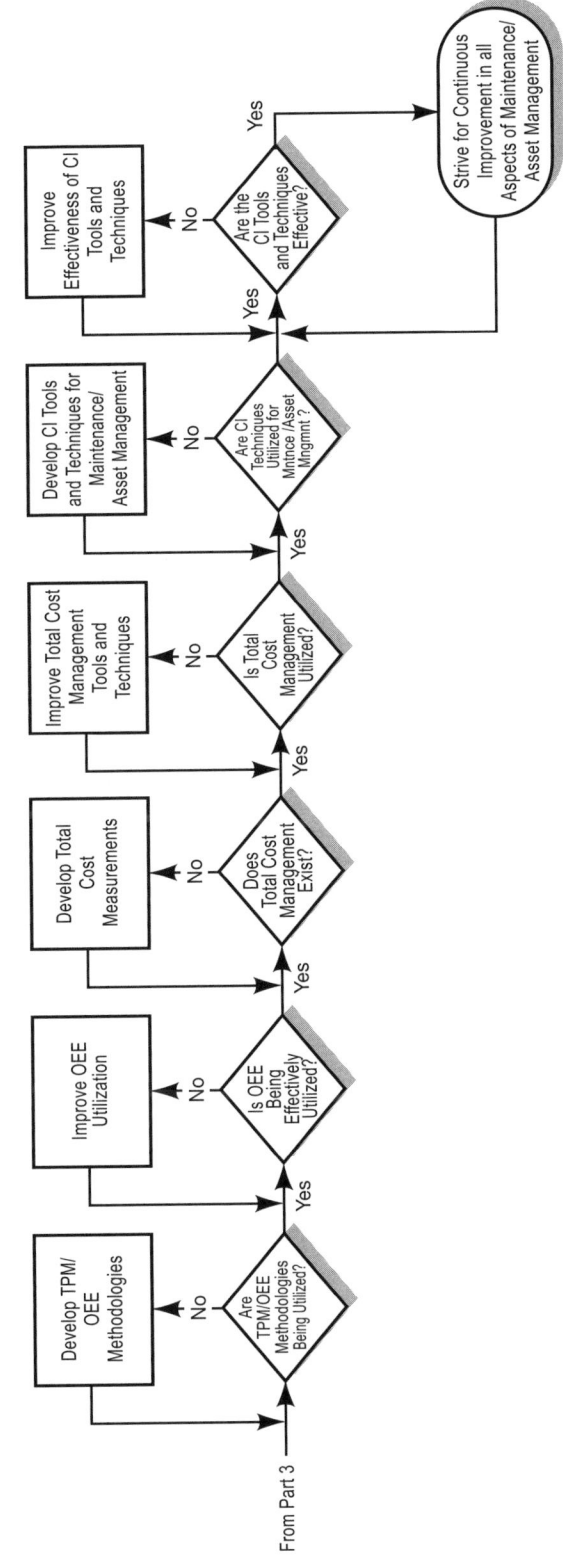

1

THE CURRENT CONDITION OF TECHNICAL SKILLS IN MAINTENANCE ORGANIZATIONS

The Perfect Storm

The term *the perfect storm* was probably applied to the technical workforce first by Robert Williamson of Strategic Work Systems (www.swspitcrew.com) in the 2005–2006 timeframe. The expression is derived from the book and movie about the *Andrea Gail*, a fishing boat doomed by a tempest while out at sea. Williamson's application has grown even more poignant since he first used it. The reason is that despite overwhelming evidence immediate action is needed, relatively few companies have taken any steps to mitigate the severity of the oncoming storm threatening their workforce. So, in weather terms, consider this chapter the Severe Storm Warning — not a storm watch. The impending perfect storm is on the horizon, clearly visible for all to see.

In the real life event, a series of weather events produced a perfect storm with ten-story waves and winds of 120 miles per hour. Similarly, a series of events are combining to produce a perfect storm that threatens the technical workforce (Figure 1.1). These events are:

1. An Aging Workforce
2. A Compromised Educational System
3. The Lack of Technical Apprenticeships

We will consider each of these "storm" elements.

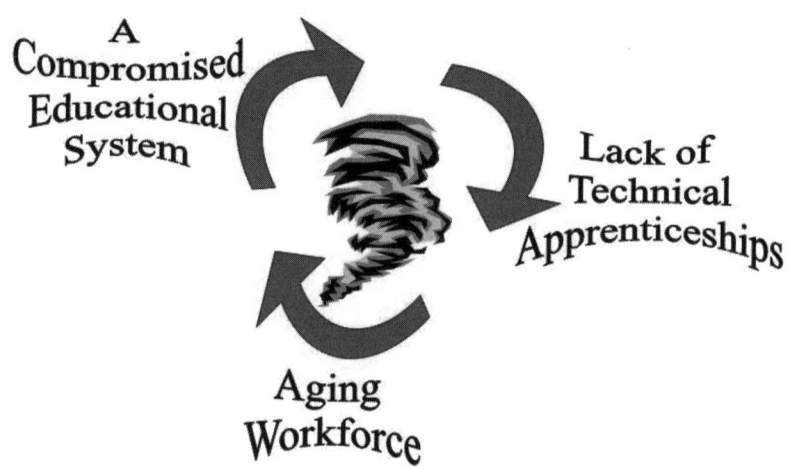

Figure 1-1 – The Perfect Storm

Author's Note

In this chapter, I will make extensive use of quotes from articles and books to document the existence of each of these elements. They are not intended to overwhelm the reader, but rather to produce conclusive evidence that the "Perfect Storm" exists. I hope that by using the information in this chapter, readers can convince their senior management team to implement at least some, if not all, of the processes that comprise the rest of the book. In this way, their company can build a "storm shelter" to insure their company's survival into the future.

An Aging Workforce

In his book *The 2010 Meltdown,* Edward Gordon writes,

By 2010 an unprecedented number of baby boomers will begin retiring in the United States, Europe, and Japan. We face the sobering reality of a smaller next generation of workers who are less well educated and less prepared with the specialized skills to run a high-tech economy.

To highlight the actual numbers, Gordon continues,

About 70 million baby boomers, some highly skilled, will exit the workforce over the next 18 years, with only 40 million workers coming in. The lack of skilled workers in technology and other sectors that face labor shortages could have a devastating effect on a company's ability to compete in the increasingly global marketplace.

The pure demographics of the workforces show that there will be problems filling positions within companies. The competition for talented, trained workers will be even more intense. In fact, Marius Leibold and Sven Voelpel, in their book *Managing the Aging Workforce*, make the following point:

> *One of the main results is a skills shortage entailing a global war for talent — in a business world primarily characterized by a concurrent cost, productivity, and quality pressures.*

So the skills shortage is not just affecting companies in the United States; it will be a GLOBAL problem. Highlighting this problem is this observation from *Workforce Wake-up Call*, edited by Robert Gandossy, Elissa Tucker, and Nidhi Verma,

> *In the United States alone, it is estimated there will be 10 million more jobs than workers by the year 2010. In other developed nations, the impact will be felt even sooner and much more intensely, due in part to highly restrictive immigration policies. By the time, this trend hits bottom, it is estimated that members of the Organization for Economic Cooperation and Development (OECD) will have suffered a combined reduction in their working age population of 65 million people.*

This text points to some relief in pure numbers, when Generation Y (born 1980–2000) fully enters the workforce by 2016, with a count of almost 73 million.

However, the text points out another problem when it notes that the problem is not as simple as replacing one set of workers with a younger generation. Even though the new workers may be technically skilled, they don't have the body of knowledge of the retiring workers. Furthermore, the number of graduates from U.S. colleges anticipated over the next ten years is short of the number needed — by 7 million workers! Complicating the matter even more is the cutback in training programs found in U.S. corporations. Therefore, *"These skill shortages will create an intense global competition for the most valuable talent."*

Many industry-specific conferences are concentrating on the aging workforce. For example, writing for the Houston Chronicle, Brett Clanton reports that the oil and gas industry "still hasn't solved what may be its biggest challenge in coming years: finding enough skilled workers." The issue has "been a central topic this week at the 2008 Offshore Technology Conference, one of the world's largest gatherings of oil and gas industry professionals." Clanton cites Thierry Pilenko, chairman and CEO of Technip, a French engineering and construction firm, who said that "projects already have been delayed because of the labor shortage, and the need for workers is only increasing as projects get bigger, more complex and more labor-intensive." According to Pilenko, "the industry [must] take more aggressive steps to address the labor shortage."

In addition, the article reports that Abe Palaz, Halliburton's director of educational and R&D partnership, advocated "a multipronged approach to tackling the issue." The oil industry "must increase and improve the talent pool in the West," he said, "where engineering talent is of high quality but in short supply." In addition, it "should move faster to develop younger employees once they are in the system and focus more on retention."

Even at a state level, this problem is acknowledged. For example, in the Bangor (Maine) Daily News, Rich Hewitt reported that Maine businesses may have trouble filling positions in the future. His article projected a high demand for skilled workers throughout Maine in the coming years. The article continued, however, that there was some question as to whether there would be enough workers to meet those needs. The accelerated retirement by many baby boomers was going to delete the talent pool, especially for skilled jobs in business, service, and manufacturing. Complicating the problem is the fact that most organizations flatten their workforces and remove redundancies, creating the skilled labor shortage. Therefore, many companies do not have mentoring or job training to move employees into the jobs created through retirements. Thus, the talent pool is not being refilled.

The same article listed high demand occupations —including electricians, carpenters, maintenance and repair workers, automobile mechanics, and waste treatment plant operators —that require technical skills. Hewitt noted the demand is not only in business and health services, where the State of Maine had focused its efforts, but in technical and craft occupations that hadn't been paid a lot of attention, such as welders, plumbers, carpenters, and the people who build Maine. "The challenge is finding someone to do Maine's work," Hewitt wrote, "We have the jobs; we can't find the people."

How are some governmental agencies suggesting that the problem be addressed? One answer comes from Oklahoma. Francis Smith reports in the Journal Record that a "bill providing tax credits to attract and retain skilled workers in target industries — aerospace, energy, advanced manufacturing and processing, biotechnology, information technology and healthcare — made it" through Oklahoma's House of Representatives recently. House Bill 3114 "would create a new law, the Oklahoma Workforce Incentives Act of 2008." Under the law, "specified workers who are employed for the first time in Oklahoma in the targeted industries within five years of graduating from an institution of higher education or training program would be granted a tax credit based on their salary." Aeronautics Commission Director Victor Bird "has been pushing for new tools to attract and retain engineers in Oklahoma." Meanwhile, the Governor's Council for Workforce and Economic Development estimated that by 2014, the state will be unable to provide the amount of trained aerospace engineers the industry demands. HB3114 is also a priority item for the Tulsa Metro Chamber, which is focused on building its aerospace industry.

The article continues, "'Oklahoma universities produced 800 new engineers last year,' said state Secretary of Commerce Natalie Shirley, 'but too many of those graduates are leaving the state to find jobs, even as engineer positions in Oklahoma go unfilled.'"

The department has begun an initiative to contact graduates who leave Oklahoma, trying to lure them back with postcards directing the recipient to a new web site that lists job openings and salaries in the state. The web site demonstrates how an engineer's salary in another state compares to the quality of life provided by the low cost of living in Oklahoma.

This problem of an aging workforce is not isolated to just a few states. A recent story from CBS News in St. Charles, Illinois asked, "In the sinking economy, where can Americans turn for a job that pays well?" The answer was, "Manufacturers are looking for employees with skills." One company even stated that it had all the orders it could fill; what it couldn't fill was jobs. The story continued, "With half the nation's 14 million manufacturing workers nearing retirement, 90% of America's manufacturers say they are short qualified workers."

Manufacturers noted that low-skill jobs were going overseas, but that better paying, high-skilled, high-tech jobs were staying. But, as they added, "many kids coming out of high school either aren't good enough at math and science or they are not interested." One high school student even commented, "I figured that was a kind of out-of-date career path." The story concluded that industry was working to update both its image and the curriculum in the schools.

This problem is not just isolated to the United States — it is a worldwide issue. A January 2006 paper published by American Society of Training and Development (ASTD) stated "Organizations around the world are faced with an aging (and sometimes declining) workforce. And they are operating in a crisis mode because of it."

The article, *Beware of the Boomer Brain Drain,* continued, "Organizations are rushing to capture the knowledge of key employees before the retirement floodgates open. But they must first determine where they're at risk, and then develop a strategy for transferring that expertise."

The article also focused on getting measureable results. It noted that the traditional measures of return on investment and return on expectations were not sufficient. A new performance measure was emerging — Return on Experience or ROX — to capture the experience of boomers who were retiring. The article also highlighted the important of mentor programs that would couple workers across generations for training purposes.

So what is the impact on companies if they take no action? A June 2008 article in Manufacturing Business Technology (MBT), reported "A new survey commissioned by Advanced Technology Services and Nielsen Research shows Baby Boomer retirement coupled with a lost generation of factory workers is creating a perfect storm, aggravating a costly skilled labor shortage for manufacturers in the U.S."

The article, *Bye-bye Baby Boomers,* continued by quantifying the losses, observing, "The need to replace these lost skilled workers has grown from a concern to a wholesale crisis in just three short years, according to the 100 senior manufacturing executives surveyed. They say the shortfall will cost their companies an average $52 million each, and even more, $100 million, for the nation's largest companies that report more than $1 billion in annual revenue."

MBT's benchmark survey polled 100 senior manufacturing executives representing companies with revenue between $10 million and $1 billion. Eighty-one percent of respondents said they would be affected by the skilled worker shortage, versus 68% three years ago, demonstrating this issue has become of even broader concern to manufacturing executives.

The 2010 Meltdown highlighted an even greater drain on profitability, suggesting that businesses could lose $60 billion per year as a result of lost productivity caused by the lack of technical knowledge.

From these sources, two facts are clear:

1. Due to the exodus of the Baby Boomer generation, there will soon be a not only a skills shortage, but also an employee shortage.
2. It is clear that companies are going to incur a considerable drain on their profits if the issue of the aging workforce is not addressed quickly.

A recent book, *The Workforce Crisis* provides a fitting conclusion to this topic:

> *To summarize, we will have too few young workforce entrants to replace the labor, skills, and talent of boomer retirees. The more immediate loss of skills and experience is already threatening the performance of many corporations. Since the generation after the boomers is much smaller, companies can no longer rely upon a relative profusion of younger workers. Even when they successfully hire and retain young workers, they are still trading experience for inexperience.*

Thus, the aging workforce presents the first element for the perfect storm: not enough new skilled workers to replace those who are leaving the workforce. Particular knowledge management objectives for the aging workforce include mining the knowledge of aging workers through an appropriate enterprise-knowledge management system. Also included is retaining all enterprise-valuable knowledge.

A Compromised Educational System

In addition to the Baby Boomers retiring, the generation positioned to replace them has been failed by the current educational systems in the United States. This generation has not been provided with the basic skills necessary to compete globally or even locally for jobs. Their basic reading, writing, mathematics, and science skills are woefully inadequate.

An article in USA Today cited a study commissioned by America's

Promise Alliance, which noted "the likelihood that a ninth grader in one of the nation's biggest cities will clutch a diploma four years later, amounts to a coin toss — not much better than a 50–50 chance." The same article also quoted U.S. Department of Education Secretary Margaret Spellings, who called "the gap unacceptable, especially now that 90% of our fastest-growing jobs require education or training beyond high school."

Consider these findings from The 2010 Meltdown, referenced previously:

- At best, only one third of all U.S. students are at the 12th grade reading level when they graduate from high school and up to 50% of current students drop out of high school.
- Recently, at Daimler Chrysler's Detroit Michigan car plants, only one out of four applicants could pass a test requiring 10th-grade skills.
- 45 million U.S. adults read below eighth-grade level.
- 50% of U.S. manufacturers found that their current workers had serious reading, writing, and math skill problems.
- Kentucky ranks last in the United States and the number of people over 25 who have graduated from high school was just 33%.

Let's consider the impact of these statistics. If we are concerned about the size of the replacement workforce, consider how the size diminishes even more when we factor in qualifications. If only 50% of students graduate high school, and then only a third of those students can read at the 12th-grade level, how many individuals entering the workforce have the reading skills necessary to function in technically advanced jobs?

The findings are no better at the Chrysler plant. If only one out of four applicants can pass a tenth-grade skills test, then only a very small portion of the total applicants would qualify to work in technically advanced jobs. This information is compatible with the finding that 45 million U.S. adults read below the eighth-grade level — about 25% of the adult population.

We can then understand why 50% of U.S. manufacturers found that their current workers had serious reading, writing, and math skill problems. Unless these problems are corrected before the baby boomers retire, we will be left with a workforce incapable of replacing the retiring skills.

The finding about the percent of Kentucky adults who are high school graduates shows that the education system has been severely compromised. If it is not corrected soon, it will leave U.S. industry in a non-competitive position. This is especially true when the compromise of the educational system is not a world-wide problem.

Note again, however, that worldwide industry faces a similar problem. A 2003 UNESCO study indicates 27% of adults in the world are totally illiterate and up to 3 billion of the world's adults still read so poorly they are functionally illiterate for most high-pay, high-skill jobs. Furthermore, although some of the world's educational systems may be compromised, an even greater problem is that, in many areas, they do not even exist.

Among the 29 countries with the strongest educational systems in place — comparing American students to international students — American 15-year-olds ranked 21st out of 29 countries in math skills. In science, they ranked 16th out of 29 countries. Finally, in reading exams, the 15-year-olds were ranked average. Twenty-five years after the landmark education study A Nation at Risk was published, American education remains in a state of crisis. Millions of students continue to pass through the public schools without mastering basic skills and knowledge.

It is beyond comprehension that a problem that has been growing for decades can be solved soon enough to provide sufficient relief in both numbers and skills of employees. This is especially true when considering the significant political and social changes that would be required. To compensate for the lack of qualified potential employees coming out of the public school systems, employers are going to have to take it upon themselves to make basic skills training a part of their internal training program.

Such efforts may seem to be too costly for industry to consider. Yet, U.S. businesses lose enormous productivity every year because employees lack basic reading and writing skills area. The cost of not solving the problem is too great not to consider starting these types of internal basic educational programs.

The Lack of Technical Apprenticeships

A 2001 study found that Germany, for the first time in many years, did not have enough apprentices to fill all the business occupational openings. By 2008, The Financial Times reported that Germany was affected by a "rapidly worsening skills shortage...one that the government [was] responding to with a package of measures due to go before the cabinet."

Furthermore, the cabinet was "expected to endorse a series of measures to open up Germany's closed labor market to foreign graduates in an attempt to tackle the skills shortage." Although economists and business representatives supported these measures, "trade unionists, in particular, see 'the measures' as further steps towards the creation of a global market for labor, and therefore a threat to the country's comparatively high wages." The report also indicated a decline in the number of engineering graduates, despite an increase in the overall number of university graduates.

Thus Germany, one of the premier countries for having skilled trades apprenticeships, severely lacked applicants (from within Germany) for these high paid jobs. The situation is even worse in other countries. For example, U. S. manufacturer Bison Gear and Engineering makes motors for everything from dialysis machines to ice cream makers. The company has all the orders it can fill. What it can't fill is jobs. As reported by CBS News reporter Cynthia Bowers, "We have about half a dozen openings right now," says Bison owner Rick Bullock. "The business is there, and we have the capacity to expand it." Bison isn't alone. With half of the nation's 14 million manufacturing workers nearing retirement, 90% of America's manufacturers say they are lacking qualified workers. Amazingly, these help wanted signs are going up at a time when the United States is hemorrhaging manufacturing jobs, nearly 3,000,000 since 2001.

"The jobs going overseas," manufacturers say, "are largely the low-skill assembly-line kind. Those that remain are high skilled, high tech, and high-paying — around $60,000 a year plus benefits."

"Every one of these machine tools that we use is about the price of a Ferrari," Bullock said. "So you need to have good computer skills, if you're going to work in high tech manufacturing today." But many recent high school graduates aren't good enough in math and science, or they aren't interested.

College professor Mark Meyer observes that lack of a trained labor force is the reason many companies move, not lack of cheap labor.

We have to ask ourselves whether the image of manufacturing has been so damaged that potential candidates won't even consider a skilled trade apprenticeship, if it were available. We then need to address what can be done to remedy this situation.

However, companies are allowing the problem to grow rather than work on changing the image of manufacturing. They should be developing and improving apprentice programs, and then selling them to potential high school candidates. Instead, manufacturers tend to take the "Chicken Little" approach to the problem — that is talk about it, but never take any action to correct it. And thus the problem continues to grow. For example:

The Orlando Business Journal, published a manpower survey, *Engineering Hardest Field to Staff, noting that,*

> *Engineering jobs have been the most difficult positions for employers to fill in this country in 2008. . . They're followed by: machinist/machine operators; skilled trades; technicians; sales representatives; accounting/finance staff; mechanics; laborers; IT staff; and production operators, according to the employment serv-ices company.*

The news release cited John Prising, president of *manpower North America,* "From our research, it is clear that across the country employers are experiencing a mismatch between the talent their businesses need and the skills and abilities, potential employees possess."

Although engineers were not on the top 10 hard-to-fill list in 2007, they were number two in 2006, the first year of the sampling. Manpower surveyed 2000 American employers this year to come up with the find-ings. Again, the image of skilled jobs related to manufacturing and process businesses is so tarnished, the majority of potential candidates are looking elsewhere, leaving many high paying jobs unfilled, which further diminishes a company's ability to compete.

Gordon's *The 2010 Meltdown* notes this shortage clearly extends into service industries, such as automobile mechanics. Gordon writes, "The nation's auto industry faces the same critical shortage of skilled technicians. According to a 2002, Wirthlin worldwide survey, only about 2% of young adults aspire to careers in the auto industry. Parents in par-ticular are likely to believe that employment as a mechanic is not enough of an intellectual challenge. Entry-level salaries of $40,000 annually or greater are going unfilled."

"The day of the grease monkey is dead," says Jim Willingham, chairman of Automotive Retailing Today. But these high-paying jobs ($70,000-$100,000 a year for master mechanic) go wanting because most people don't understand that the industry has changed drastically.

The president of Siemens observes that education must be the key to the standard of living, not location.

It is again clear that the potential of filling these high-paying jobs is being severely impacted by a disconnect between what manufacturer's have to offer and what the educational system perceives as the future of high-skilled jobs in manufacturing. Unless this disconnect is eliminated, we are in jeopardy of losing the ability to manufacture products and fully utilize our asset investments. Educational institutions need to communicate this information to their students. Manufacturers are going to have to provide the apprentice programs to produce highly-skilled technicians capable of helping to maximize a company's investment in their asset (equipment) base.

Although this may seem to be a change that is out of reach for many companies, it is not optional. If companies want to survive, they must make the changes. As Leibold and Voelpel observe,

> *Now, however, the world is entering a new era of unavoidable cost escalations, especially due to the significantly increasing cost of skilled human resources. It is a simple issue of supply and demand: the shrinking supply of talent and skills and the increasing demand for high level, knowledgable, and innovative labor forces, leading to workforce cost escalations.*

In other words, technicians in manufacturing are going to become increasingly expensive. This picture is opposite to the one commonly portrayed by high school counselors who push their students toward college rather than helping them find quality apprenticeships in which to enroll. Most quality apprentices will have a high standard quality of life as they progress toward journeyman status in their particular skilled trade.

Liebold and Voelpel also write, "It also requires a new mindset concerning employees and their related cost: in the future, enterprises will see their workforce increasingly as an investment, rather than just cost to be contained or controlled."

Thus, on-going training for their employees will be a cost of doing business for most companies. Those companies that fail to invest in additional training for their skilled employees will likely lose the employees to more progressive companies.

This point is further emphasized by Gandossy, Tucker, and Verma, when they note,

> *There is an implicit idea that hours equal output and that the more hours people put in, the more they — and their organizations — accomplish. This concept dates back to the days of agricultural production, and the only way to plow more was to work in the fields*

longer. It carried on through the days of physical production. When there was closer connection between hours and output. These days, however, the connection between hours and performance is much less clear, because increasingly we are dealing with creativity, innovation, and intellectual work, rather than sheer physical labor.

Traditionally we have measured an employee's contribution by some form of visible output. However, this approach is not truly the case today and will become even less applicable in the near future. Furthermore, the relationship between employee and financial performances is a complicated one. Each builds on the other. Strong employee performances improve financial performances. At the same time, strong financial performances encourage investment in human resources in a way that enables the organization to attract and hire the best employees. This circular and reinforcing relationship reflects what we refer to as the success spiral. Companies that invest in talent have the potential to generate financial successes, which in turn generates funds to further invest in talent, hence continuing to build on their success.

Conversely, the death spiral refers to situations where companies respond to market downturns or other financial pressures by cutting back on people investments. This results in the departure of pivotal employees and reinforces the likelihood of future poor performance and financial distress."

During a slowdown or even an economic recession, what do most companies do? Unfortunately they begin to downsize the workforce, which leads them to the death spiral. Traditional business schools teach that in downturns you reduce your expenses, including the workforce. However, when business begins to increase, the talented employees will have found positions in other, more progressive thinking companies. These will be companies that are riding their success spirals. Executives will need to determine what type of company they want to run — what type of business spiral they value.

Why do we have this conflict? Much of it is due to what is being taught in most business schools today. Almost exclusively, the schools teach their student to focus on the quarter-to-quarter bottom line, with few thoughts to strategic 3–5-year planning and investment. Few schools focus on one of the most important costs they can consider: the cost of knowledge. Workforce Wake-up Call noted,

The cost of knowledge is largely hidden: for example, what is the

cost of delaying the introduction of a new product that was being worked on by a recently retired scientist? And what is the cost of reduced efficiency or increased errors in a plant where two first-line supervisors just retired? Answering questions like these can provide a starting point for investigations and to provide solutions when faced with such realities.

Companies that do understand the value that departing employees take with them when they retire from or leave the company will never invest enough in planning for their replacement.

Applying these perspectives directly to the maintenance / reliability disciplines, the ARC Advisory Group report *Best Practices for Maintenance Management* showed that an aging workforce has become an issue in nearly all industrialized countries due to the baby boomers approaching retirement without enough younger people to replace them. Their findings showed that virtually everyone shared these concerns. Participants in the report were primarily concerned with knowledge and skill retention. The report also noted that younger people preferred other professions, making it especially difficult to recruit replacements. Finding qualified replacements was a key issue for everyone.

Companies must come to treasure training and development — employers must not only encourage employees to continue seeking education, but also provide that education directly in order to maintain needed skill levels.

The 2008 – 2009 Economy

In midyear 2008, the economy world-wide slowed down. By the 4th quarter, the slowdown turned into a recession. This recession led to companies cutting costs; world-wide, millions of jobs were lost. For companies that focused on cost cutting, two major areas of focus were maintenance and training expenses. Travel and educational budgets were reduced, if not eliminated.

How does this affect the perfect storm? On the surface, it may seem the job reductions would make more skilled trade labor available to delay the onset of the perfect storm. However, this was not the case. Certain industries, particularly the automotive industry, offered massive programs of early retirement and employee buy-outs, which simply accelerated the "Aging Workforce" element of the perfect storm.

In a slow economic time, there are great advantages for companies that can optimize their asset utilization so they can produce the maximum volume of their products, in the shortest time, with minimum waste. However, companies are currently not taking this approach. Instead they are reducing their capacity by closing factories and eliminating jobs. Although it is beyond the scope of this text to detail all of the factors that led up to the current recession, it is clear that most companies still do not understand that maximizing their asset utilization would have minimized the impact that the recession is having on their business.

If companies had focused on optimum utilization of their assets, instead of building over-capitalized plants, the amount of excess manufacturing capacity would have been reduced. The impact of the business downturn would have been minimized and the recession may not have occurred. With this being said, highly skilled maintenance and operations technicians are critical to maximizing a company's asset capacity. It is clear that, currently, companies have not yet realized this fact. Whether they realize this when the recession eventually begins to lessen — and rebuild their technical apprentice training programs and restock their maintenance and operations organizations with the highly-skilled individuals that will be required in the immediate future — only time will tell.

If we were to use a NASCAR analogy, the economy is currently running under a yellow flag, and companies have a choice. They can stop in the pits and prepare their car for the green flag restart, by adding fuel, changing tires and making any chassis (organizational) adjustments necessary to optimize the car's performance. If they choose to stay out on the track and not take advantage of the current conditions to prepare for the green flag restart, they will likely be quickly passed by competitors that did prepare, once the green flag is waved.

The economy worldwide will recover, that much is beyond any doubt. Many companies are not focusing on the future, but rather on the immediate situation. They are making some short-term decisions that will cripple their competitiveness in the near future. In fact, some companies may weather the recession, but go out of business as the economy recovers, because they are focusing on and making wrong short-term decisions.

I sincerely hope that some of the material in this textbook (and others as well) can be used as a catalyst to help stimulate the thinking of senior executives who will need to make these types of decisions.

Concluding Thoughts

The Perfect Storm is real. This chapter has presented clear evidence that the storm consists of three major factors:

1. An Aging Workforce
2. A Compromised Educational System
3. A Lack of Technical Apprenticeships

These factors have come together to create a massive challenge to all companies today. The issues can no longer be ignored. They are here now and a reality. These issues are ones that all companies must address in the next decade or face extinction.

We can count on three trends to continue to develop. They are:

- **Shortages of skills and experience.** As the baby boom genera
tion reaches retirement age, organizations face a potentially debil
itating brain drain of skills and experience very.
- **Shortages of workers.** Overall demand for workers is already
beginning to exceed supply. The gap is projected to grow mi-
lions, perhaps tens of billions, of workers, with potentially pro-
found effects on economic output and standard of living area.
- **Shortages of educated candidates.** Despite continuing progress
in average educational achievement, colleges will graduate too
few candidates to fill the technical, information intensive, judgment intensive, jobs five years from now.

In the remainder of this text we will present a plan of how to make sure employees have:

1. Education that assures a firm educational foundation
2. Specialized technical skills education and training
3. Continuing technical education and training throughout their working lives to keep them up to date.

The clear goal is that companies that apply the principles in this plan will survive their "Perfect Storm."

Gordon, Edward E. The 2010 Meltdown: Solving the Impending Jobs Crisis. Praeger, 2005.

References

Leibold, Marius and Sven Voelpel. Managing the Aging Workforce. John Wiley and Sons, 2006.

Gandossy, Robert et al. Workforce Wake-up Call. John Wiley and Sons, 2006.

Clanton, Brett. The Houston Chronicle, May 9, 2008.

Hewitt, Rich. State Says Workers Needed. Bangor Daily News. April 24, 2008.

Smith, Francis. The Journal Record. Oklahoma, March 12, 2008.

CBS News. "Building Jobs — in Manufacturing." St. Charles, Illinois, March 11, 2008.

American Society of Training and Development (ASTD). Beware of the Boomer Brain Drain. January 2006.

Manufacturing Business Technology. Bye-bye Baby Boomers. June 11, 2008.

Dychtwald, Ken, et al. The Workforce Crisis. Harvard Business School Press, 2006.

Toppo, Greg. Crisis graduation gap found between cities and suburbs. USA Today; August 1, 2008.

Benoit, Bertrand. "Germany seeking skills warms up its welcome." The Financial Times. July 14, 2008.

Bowers, Cynthia. "Building Jobs in Manufacturing." CBS News. St. Charles, Illinois. March 11, 2008.

Manpower Survey: Engineering Hardest Field to Staff. Orlando Business Journal. April 21, 2008.

ARC Advisory Group. "Best Practices for Maintenance Management" August, 2007. www.arcweb.com

THE TRAINING DEVELOPMENT PROCESS

In this chapter, we will discuss a process flow for developing training programs. In many companies, training programs were developed because of a perceived need for them. In many cases, however, training was not the answer. It may have been organizational issues, past practices, or some other factor that made situations appear that they were caused by a lack of training for the individuals involved. (This point will be discussed in Chapter 11 when Analyzing Performance Problems is discussed.)

The process in this chapter also avoids the "Sheep Dip" type of training programs. The "Sheep Dip" program is the type of training where you bring everyone into a room, present the material, and then rush them out the door, and hope some of it sticks. This type of training program will never produce the type of performance improvement that is required to justify and sustain a technical training program.

Developing the Training Program

The process flow for developing training programs is highlighted in Figure 2-1 on the following page.

Defining the Need

The first step is to understand why a training program is actually needed. If it is to develop new technicians, then there are three types of training programs that will need to be developed. The first is a basic skills program that insures that new employees have the basic technical job skills necessary to perform in their new role. The second training program is a formal apprentice training program that will help employees progress

```
┌─────────────┐   ┌──────────────┐      ◇ Developing ◇        ┌───────────────┐
│  Technical  │   │   Lack of    │      ◇ Technical Skills ◇  │    Need to    │
│  Training   │   │ Replacement  │◄─────◇   Training?     ◇──►│ Upgrade Current│
│ Development │   │ Skilled Labor│      ◇               ◇     │ Technical Skills│
│Process Flow │   └──────────────┘      ◇             ◇       └───────────────┘
└─────────────┘                              │
                                             ▼
                                      ┌──────────────┐
                                      │ Conduct Duty │
                                      │   Analysis   │
                                      └──────────────┘
                                             │
                                             ▼
                                      ┌──────────────┐
                                      │ Develop Task │
                                      │   Analysis   │
                                      └──────────────┘
                                             │
                                             ▼
                                      ┌──────────────┐
                                      │ Develop Need │
                                      │   Analysis   │
                                      └──────────────┘
                                             │
                                             ▼
                                      ┌──────────────┐
                                      │  Duty-Task-  │
                                      │     Need     │
                                      │  Validation  │
                                      └──────────────┘
                                             │
                                             ▼
                                      ┌──────────────┐
                                      │   Develop    │
                                      │ Instructional│
                                      │  Objectives  │
                                      └──────────────┘
                                             │
                                             ▼
                                      ┌──────────────┐
                                      │Develop Course│
                                      │  Materials / │
                                      │   Identify   │
                                      │  Resources   │
                                      └──────────────┘
                                             │
                                             ▼
                                      ┌──────────────┐
                                      │    Select    │
                                      │ Instructional│
                                      │   Methods/   │
                                      │   Trainers   │
                                      └──────────────┘
                                             │
                                             ▼
                                      ┌──────────────┐
                                      │   Develop    │
                                      │  Evaluation/ │
                                      │  Assessment  │
                                      │   Methods    │
                                      └──────────────┘
                                             │
                                             ▼
                                      ┌──────────────┐
                                      │ Develop OJT  │
                                      │    Tasks     │
                                      └──────────────┘
                                             │
                                             ▼
                                      ┌──────────────┐
                                      │   Deliver    │
                                      │   Training   │
                                      └──────────────┘
                                             │
                                             ▼
                                      ┌──────────────┐
                                      │   Evaluate/  │
                                      │    Improve   │
                                      │   Training   │
                                      │   Program    │
                                      └──────────────┘
                                             │
                                             ▼
                                      ┌──────────────┐
                                      │  Reevaluate  │
                                      │ Training When│
                                      │ Any Changes  │
                                      │   are Made   │
                                      └──────────────┘
```

from beginners to journeyman-level technicians. The third is an ongoing skills enhancement program, which allows journeyman technicians to continue to upgrades their skills as new equipment is brought into a plant or new maintenance tools and techniques become available. All three of these types of training programs will be discussed in depth in Chapter 3.

The first step then is to define what is driving the training initiative. If it is a lack of currently qualified employees to fill openings in the current organization due to attrition, then it is likely all three types of training will need to be developed. If the training initiative is being driven by the need to upgrade current technical skills of existing employees or if there has been equipment added to the plant, then it is quite likely that only the ongoing skills enhancement program will need to be developed or modified.

Duty, Task, and Needs Analyses

When it's decided that a technical training program needs to be developed, the next step is to conduct a duty (or needs) analysis.

Note: No individuals in an organization have the knowledge or experience to perform a duty analysis on their own. It is necessary to identify subject matter experts (SMEs) who will participate in the duty analysis. They may include supervisors, senior technicians, and planners.

If the training program is being designed to add qualified employees to the organization, the duty analysis will examine all of the duties that a qualified employee should be able to perform in that job role. If the training program is being designed to enhance the skills of existing technical employees, then the duty analysis will need to focus only on the newer duties that the employees are being asked to perform. If the training program is being designed due to new equipment being added to the plant, the duty analysis will need to focus only on the new duties that the technical employees are being asked to perform on the new equipment.

After the duty analysis is completed, it's necessary to perform the task analysis. This analysis evaluates each duty and details each step (or task) necessary to perform the duty. Again, this analysis is typically performed by interviewing the subject matter experts who participated in the duty analysis. In some advanced maintenance organizations, a detailed plan for the particular duty may provide a suitable outline for the task analysis.

After the duty and task analyses are completed, each task is examined for the knowledge necessary to perform the task. The amount of information that is gathered during the need analysis stage is usually voluminous. It is best to design an electronic database that can be used to collect and store this information. Once the needs analysis is complete, the needs must be sorted and common elements combined to eliminate redundancy. For example, suppose an apprentice has been trained on the proper and safe use of hammers. Then, when being trained on how to properly and safely use a chisel, the hammer instruction would not need to be repeated, but only referenced.

Validation

Once the duty, task, and needs analyses are completed, it is necessary to validate the entire analysis. Validation is typically performed by a detailed review by the subject matter experts who have been involved in the previous analyses. Although this may seem to be a laborious task, validation is critical to ensure the development of an effective training program.

Develop Objectives and Materials

Once the validation process is complete, instructional objectives can be developed. With the previous analyses completed, the material is divided into logical topics such as hand tools, power tools, hydraulics, and pneumatics. These topics are further divided into subject outlines. For example, if using hand tools is the topic, subject outlines might include hammers, chisels, punches, wrenches, sockets, and ratchets. The instructional objectives detail what is to be accomplished in each of the subject outlines being taught. This process allows proper evaluation of the student (and the instructional material) at the end of the lesson.

When the materials that are to be taught are clearly identified, the next step is to develop the training materials or identify available training resources that cover the topics that will meet the instructional objectives. In some cases, standardized materials available off the shelf cover the topics identified. In other cases, instructional materials must be developed based on the identified topics. These may be written in-house by writers, using subject matter experts and reference material. Another option is to identify outside experts who teach the identified topics, although this option is usually reserved for specialized equipment specific training.

Identify Methods, Trainers, and Outcomes

As the material is being prepared, it will be necessary to identify both the instructional methods that are to be utilized and the trainers who will actually present the material. The instructional methods may include lecture, group discussion, hands-on demonstrations, and laboratory exercises. Each of these instructional methods has both positive and negative aspects. It is up to the trainers to determine which method will best reach the intended audience. Furthermore, each of these instructional methods may incorporate different delivery techniques such as white boards, videos, overhead projectors, and computer projectors. In turn, each of these delivery techniques has its own positive and negative aspects. The trainer must select the techniques most appropriate to deliver the material to the trainees.

Identifying the technical trainers who will actually present the material is another challenge. Technician and apprentice trainees who attend these types of sessions have a tendency to respect instructors who have actually been involved in the technical skilled trades themselves. They respect the real-world experience and the knowledge that journey-man trainers possess. Previous on-the-job experience indicates that the trainers will have good technical skills. In addition to strong technical skills, the trainers also need good presentation skills.

With the courses clearly defined and developed, the measurement of the outcome of the training must be clearly defined. The instructional objectives have already been specified in a previous step. Therefore, this step should define how to measure whether those instructional objectives have been achieved and to the degree they have been achieved. Measurement can be made by a written test with true-false, multiple choice, or essay-type questions. It can also involve a demonstration of the learning in a lab setting, with the trainer observing the skills and techniques of the students. Computerized modeling techniques can also be used to evaluate certain types of training.

Any complete technical skills training program should have an on-the-job (OJT) component, where the students can see the actual application of the course material and how it can be used on their specific job assignments. This component requires coordination between the classroom or lab trainers and the on-the-job instructors. The OJT instructors must always know what classroom training the students are receiving so it can be matched to the appropriate OJT assignment. For example, if the "academic" portion of the training covers operation of overhead door circuits, it would be beneficial for OJT instructors to find a similar electrical control circuit for the OJT training component. The trainees could then quickly relate the classroom training to the actual performance of an equipment item they encounter on their job. This theory–application type of training completes the cycle of learning and applying the information.

Delivering and Evaluating the Program

With the delivery methods, trainers, and evaluations in place, the training program is complete and ready to deliver to the designated students. Some organizations will run a test class for the subject matter experts who have been involved in the training program development. This test class will identify any final material changes or adjustments that need to be made before the class is presented to the students. Other organizations believe they have performed their due diligence during the development process and move into direct presentations of the training material to the students.

As students begin progressing through the training program, the materials in the program can be evaluated. If the training program achieves the objectives that were detailed at the start of the program development, then it can be considered a success. If, however, the instructors, the students, the trainee's supervisors, or even the OJT instructor identifies areas for improvement, these improvements should be incorporated in the training program as soon as practical. There are always new materials, new delivery methods, or ideas that can be incorporated to continuously improve the training programs.

In time, all things change, whether it is the job descriptions, the entry level skills of the employees, or the plant equipment. Many factors such as employee turnover, new equipment purchases, etc., may require that a training program to be reevaluated. When the occasion arises, the same level of effort that was put into developing the original program should be applied to changing and updating the training program. In this way, the training program becomes a "living" program and will always meet the needs of the technical organization.

This chapter provides a basic outline of the training development process. Subsequent chapters will present these topics in greater detail, allowing you to successfully develop a technical training program for your organization.

3

Duty – Task – Needs Analysis

Introduction

In this chapter we will discuss what is required to perform duty–task–needs analysis. This process is what focuses the training program on specific topics designed to affect the skills and behaviors of the skilled-trade apprentices as they progress to skilled-trade journeymen. In order to have a clear understanding of the impact duty–task–needs analysis has on the training development process, it is necessary to understand what a comprehensive skilled-trades program involves. The following example is a case study that will provide a foundation for the remainder of the material in this text. It is not to say that this process is the only one that can be used to develop skill trades training, but it is a model that encompasses the main components of a successful skills training program.

Skilled Trades Training Program Case Study

This case study is from the primary metals industry. The plant had 7500 employees of whom approximately were 1500 skilled-trade employees. The maintenance organization was deployed in an area configuration with central shop support. The shops had specialized skills such as welders, riggers, machinists, pipefitters, and electrical rebuild. The area maintenance technicians were super crafted in that they were expected to be proficient electrically, mechanically, and in fluid power. The following material will detail the training program that was used to produce the skill levels necessary to maintain the company's assets.

Why use this company as a case study? The maintenance organization had the skills and processes necessary to win the NAME (North

American Maintenance Excellence award). Details on the award can be found at www.nameaward.com . The focus of this material is not to win the award, but to show how to develop a skills training program that can support the skills need of a "Best Practice" maintenance organization.

1. Introduction to Maintenance Skills Trades (Trade and Craft Orientation)

This program was developed as a way to pre-qualify candidates for entry level positions in the skilled trades. In addition, the trainees were given instruction to provide them with the skills necessary to be useful as helpers when transferred into a specific skilled trade department. The trainees were pre-selected by the Human Resources department, so they were not given a pre-qualification skills test. The only basic requirement was a high school education or a GED.

The program was a two-week class for 40 hours per week. The 80 hours total included classroom training, shop training, and safety training. The classroom training started with a basic math refresher, including fractions, decimals, basic algebra, and some trigonometry. Why this level of math? It was to support the other academic class, which was blueprint reading. Therefore, the math class had to support adding and subtracting fractions and decimals, because these math skills are required to read almost any blueprint. The blueprint class started with basic shape identification and progressed through to being able to read standard shop blueprints for the plant. These two classes accounted for 20 hours of class time.

The math and blueprint classes led to shop layout projects. Using the information that was gained in the two classes, the students were required to lay out a 16" x 16" x 1/2" steel plate into four equal quadrants. Then they were required to lay out a circle, a triangle, a square, and a rectangle in each of the quadrants. The object had to be centered in each of the quadrants. The layout of the steel plate is pictured in Figure 3-1.

In addition, the student had to lay out eight 12" x 2" x 1/2" steel bars after cutting them on a band saw from an 8-foot piece of 2" x 1/2" shop stock. They were also required to center punch eight evenly-spaced holes in each bar. The layout of the steel bar is pictured in Figure 3-2.

The students were then required to perform certain operations such as drilling and tapping holes in the bars. They did work on a drill press, with a hand drill, with a hand tap, and with ball peen hammers and rivets. They used a treading machine to tread 3/8" pipe nipples, 3/8" bolts. The

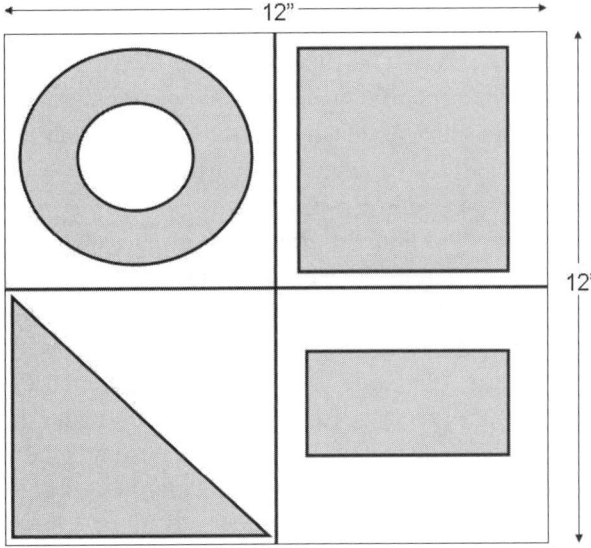

Figure 3-1 Steel plate layout

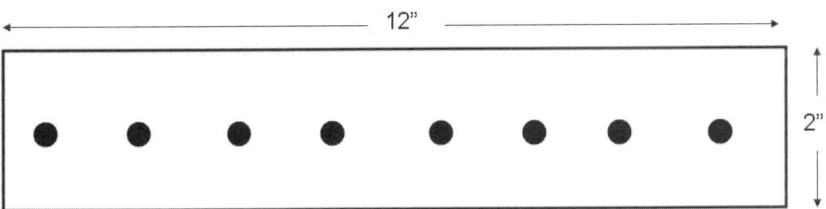

Figure 3-2 Steel bar layout

pipes and bolts were required to fit properly into the tapped holes on the bars.

After several practice sessions, the trainees were required to use an oxy-acetylene torch and a natural gas torch to cut the shapes out of the 12" by 12" square. They would then use a pedestal and hand grinder to smooth the cut edges. On some shapes they would grind bevels; on others they would have to bevel the edge with one of the torches.

Other demonstrations and class activities involved chain hoists, small pendant cranes, gear case disassembly, and ladder safety.

At the end of the two weeks, the department supervisor for each of the trainees was invited to the training lab / classroom to inspect the various projects the trainees had completed. The instructors were required to prepare a follow-up report on each student, highlighting their individual strengths and weaknesses. The supervisors in the department could then follow up and provide specific mentoring that would allow the students to prepare properly for entering the apprentice training program.

2. The Apprentice Training Program

The apprentice training program was composed of two parts: time on the job and academic proficiency. The time on the job started with an apprentice in a Zone 3 pay grade. Every 6 months, if job performance was satisfactory, the apprentice was advanced a grade until Zone 8. An apprentice could not progress to Zone 9, which was considered to be an entry level journeyman, until the academic part of the apprenticeship was satisfactorily completed. Once the academic part was completed, the progression was to Zone 9, 10, 11, and finally 12, which was the maximum pay grade for a journeyman. There was another advancement, which was for a lead. This advancement allowed progression to Zone 13, but this was a bid position that was usually claimed by the most senior person on the crew.

The academic portion of the apprenticeship was 832 hours of classroom and lab training. The usual schedule was 1 day per week for approximately a two-year time period. However, there were times when there were forecast demands for journeymen level individuals. At such times, the program was accelerated so that apprentices were in school for a week for every other week. This schedule (allowing for some skipped weeks) allowed participants to finish the academic portion of the apprenticeship in approximately six months. In a 15-year time period, this schedule was used only twice — and once was after the training had been suspended for a time period (due to a business downturn) and was used to allow the apprentices to progress to journeymen.

Because the journeyman status was for a "super craft" skill, the academic training was comprised of electrical, mechanical, and fluid power training. The division of the training hours was 416 for electrical and 416 for the mechanical and fluid power training. The electrical training was divided into AC and DC theory and labs. The apprentices were able to progress through the various types of control circuits and connect up the circuits on the lab equipment in the classroom. The AC courses covered 110 volt to 440 volt control systems. The AC lab equipment was mounted on control boards which surrounded the perimeter of the classroom. Students could literally study a circuit and then walk to the wall and connect it up. The wiring was behind the boards, which were swivel mounted so the instructor could open the back of a board and introduce faults for the apprentices to troubleshoot and diagnose.

The DC lab was set up differently, with the classroom on one side of the room and the DC controls and equipment on the other. The DC equipment included controls and motors. The courses were organized similarly, with the theory taught, and the lab exercises conducted after the appropriate classroom material was covered.

The mechanical training covered material related to bearings, belt drives, chain drives, gear drives, couplings and alignment. The courses covered theory of operation and proper maintenance procedures. Once the classroom material was covered, the apprentices would be able to handle all of the components, actually installing them in specialized lab devices and study their operation.

The hydraulics and pneumatics classrooms and labs were also similar, with the theory of operation covered in the classroom and then using quick disconnect fittings, actually connecting hydraulic and pneumatic circuits. The instructors were able to introduce faults into the circuits for troubleshooting and diagnosing problems.

The grading periods were 13 weeks (or classes) in length. The apprentices were graded on their progress. They were required to achieve a 70% or better to stay in their apprenticeship. If they fell below 70% in one grade period, they were given another grade period to raise their average. If they did not, they were removed from the apprenticeship. If they were removed from the program, they were provided a list of sources where they could take remedial training in the area they failed out of the program. Once they felt they could re-qualify, they were allowed a second attempt at the apprenticeship. If they failed a second time, they were transferred to operations and their seniority would determine what jobs they could move into for the future. However, they were not allowed to re-apply for maintenance positions in the future.

As a part of the apprentice training program, there was an in-department training component as well. This is similar to what is called On-The-Job (OJT) training. The OJT training took place during the week after testing in the grade period. All of the in-department instructors would be kept informed as to the topics that were covered during the 13-week grading period. Their responsibility would be to arrange for the apprentices to perform tasks in their department that made practical application of the material covered in the classroom setting. If possible, the in-department instructor would choose a mechanical and an electrical task. Once the apprentices had completed the tasks, with the instructor observing and coaching, the task completion was recorded with any in department instructor observations. The information was then forwarded to the training department administrator for review and recording. If any in-department instructor observations required a change in the course material, this observation was noted and passed on to the classroom instructor for further evaluation. A return reply was required to the in-department instructor.

3. On-Going and Advanced Journeyman Training

Ongoing journeyman training was typically developed when new equipment was brought in to the plant or when new maintenance techniques and tools were developed. The training was usually originated by the department receiving the new equipment. If the training was related to new maintenance techniques and tools, the request for this training could have been originated by one of the training instructors, a maintenance engineer, or anyone in a maintenance department who had knowledge of the new tool or technique. Individuals could have gained this knowledge from attending a seminar or a technical tradeshow. Regardless of the source of the origination of the request, it was typically handled by one of the training department heads, depending on whether it was mechanical or electrical.

Upon investigating the request, the training department head would survey the various departments to validate the request. If the request was valid, a short duty–task–needs analysis would be conducted. The training would then be developed using the guidelines presented in the apprentice training section. Once the training was finished, it would be offered to each department manager for evaluation for their assigned employees. The final number of employees that would be required to take the program was identified; the training would then be scheduled so that it would not conflict with the apprentice training program. After the program had been conducted, the students were required to fill out evaluations, which were used to further improve the training program.

As time progressed, the plant began to modernize much of its equipment. It became evident, that the electrical training was not sufficient for the technicians to be able to maintain and troubleshoot the more modern equipment. Plant management decided that there would be a new classification of maintenance journeyman.

This new classification —electronics technician — would be required to take training that the internal training staff was not capable of providing. After visiting one of the local universities, the training manager determined there were five classes the university was teaching in advanced electronics that could fill the need for the advanced training. Thus, a group of maintenance journeyman was selected to take these classes. The first time the classes were taught, it was not as effective as management had hoped it would be. It seemed the professor wanted to teach the students how little they knew and how much he knew. However, after the first classes were taught, and this problem became apparent, the training manager learned that one of the internal electrical training instructors had finished his bachelor's degree in electronics. The manager then renegotiated with the University for the internal instructor to teach the five electronics classes at the University. This approach was much more effective because the internal instructor could use real-world examples to develop applications of the electronics material.

Under the plant's agreement with the University, the plant was allowed to fill a certain number of seats in the class. Any seats above that number were for open enrollment. This became a win-win situation for the university and the plant. The plant always had enough employees to fill their seats in the class. The university benefited because once this arrangement became public knowledge, other plants in the area would send their journeyman technicians to these classes, knowing the level of instruction and information they would be receiving. These classes were always full, and usually had a wait list for individuals wanting to get into the classes.

This example illustrates how well industry and the educational system can work together to fill certain training needs. We will consider other examples, as we progress through this text. This model training program (as well as a few others) will be referenced as we go through the remainder of this text.

The Duty–Task–Needs Analysis

Duty–task–needs analysis has been mentioned many times thus far, but what is it? Simply stated, a duty–task–needs analysis examines what duties an individual is required to perform, what tasks must be accomplished to perform these duties, and what knowledge and skills are needed to perform each task? This approach focuses training so that it best prepares technical employees to perform their jobs properly.

Unfortunately, in many companies trying to implement worker flexibility, the exact duties required from each person becomes hazy. For this reason, Volume 3 of the Maintenance Strategy Series, Maintenance Work Management Processes, highlights the value of work processes, blended with swim lanes (see page 51 of that volume). Although it is not "politically correct" in many companies to have defined job descriptions today, such descriptions are not necessary for a duty–task–needs analysis to be successful. In Maintenance Work Management Processes, specific job descriptions with detailed roles and responsibilities were presented for supervisors, planners, and maintenance engineers. It is this style of job descriptions that are required to make the duty–task–needs analysis successful.

Author's note: Some companies will try to develop a corporate model for the training. This means they will try to develop all of the training materials at a corporate level and deliver them to the individual plants for the actual implementation. Any company attempting to do this must clearly understand the importance of standardization of the maintenance and operations job descriptions and work processes. If there are deviations in the job duties from plant to plant, then the standardized training may not apply. If one plant is unionized, with very specific job descriptions and lines of job jurisdiction, their duty-task-needs analysis will produce one type of training program. If another plant is more of an "open" shop, with more flexibility in work assignments, their duty-task-needs analysis will produce another type of training program.

The temptation for a corporate initiative will be to compromise the duty-task-needs analysis and create a hybrid training program that tries to address the needs of both plants. Unfortunately, this will compromise the development effort. In trying to blend the needs of both plants, the program will meet the needs of neither. The blended duty-task-needs analysis will produce material that will not apply in either plant, which will result in cutting and pasting the derived instructional objectives. The entire process has now become subjective, based on the perceived need of the trainees.

If companies are going to be successful in developing a standardized training system, they will need an organizational configuration similar to the one described in Volume 4 of the Maintenance Strategy Series on pages 9–13. The evolution from plant-level programs to corporate- or enterprise-level programs was explained regarding maintenance information systems. If companies have that level of organizational and process maturity, then they can be successful standardizing their technical training program. If this level of standardization can not be achieved, then it is likely the training they develop would be inconsistent, lead to excessive training costs.

How does the process begin to identify the duties of a mechanical maintenance technician (or some other technician)? It begins with a well-constructed survey, which would be given to the mechanical technicians as well as individuals who supervise or plan for the technicians. This survey can be a questionnaire that asks them to identify the typical duties they perform. Although it is not practical to survey the entire department, it is necessary to have a large enough audience to validate the survey.

Once the surveys are completed and collected, it will be necessary to consolidate the information. It will then be necessary to validate the findings. This validation can be handled in two different ways. The first is to prepare another survey with a complete list of all the duties gathered during the first survey. This approach can be time consuming, and the individuals completing the second survey have been known only to give the survey a cursory look and approve it. The second method is to use an interview process. In this process, a select number of individuals representing a good cross-section of the respondents are chosen for interviews. The individual responsible for conducting the duty analysis then schedules time with each individual and personally goes through the duty list with them. This more focused approach seems to get better results to complete the duty survey.

For example, if we were to list the duties of a mechanical maintenance technician, what would be included? The following list is a sample of some of the duties of the mechanical maintenance technician.

1. Demonstrate basic employability skills.
2. Demonstrate knowledge of general safety regulations.
3. Demonstrate knowledge and ability to read and interpret shop blueprints.
4. Demonstrate knowledge of basic shop theory.
5. Demonstrate knowledge and ability of how to use hand tools.
6. Demonstrate knowledge and ability of how to use power tools.
7. Demonstrate knowledge and ability of how to use shop equipment.

8. Demonstrate knowledge and ability of how to use gas welding and cutting equipment
9. Demonstrate knowledge of mechanical fundamentals
10. Demonstrate knowledge and ability to use lubricants properly.
11. Demonstrate knowledge and ability to install and maintain mechanical drive components properly.
12. Demonstrate knowledge and ability to install, inspect, or replace bearings properly.
13. Demonstrate knowledge and ability to maintain and repair pumps properly.
14. Demonstrate knowledge and ability to maintain and repair hydraulic system components properly.
15. Demonstrate knowledge and ability to maintain and repair pneumatic system components properly.

And this list would continue on until all the duties of the mechanical maintenance technician were fully identified.

Author's note: This list is an example only. It should not be used as a duty list for the mechanical maintenance technicians at your site. Your list should be developed from the survey process mentioned above.

Once the duties have been identified, and there may be dozens of duties, you are ready to begin identifying the tasks. The tasks basically describe the various steps necessary to complete each duty. For example, using item number 12 as a guide, how would a mechanical maintenance technician demonstrate the knowledge and ability to install, inspect, or replace bearings properly? A list of tasks might be as follows:

12.1 Identify common bearing types and their applications.

12.2 Demonstrate ability to properly mount and dismount bearings

12.3 Identify the proper bearings seals for specific applications.

12.4 Demonstrate ability to properly lubricate various bearing types.

12.5 Identify various bearing wear patterns.

12.6 Demonstrate ability to find bearings in appropriate catalogs.

Here is an observation that shows an important shortcut to the duty–task process. If the Maintenance Strategy Series has been followed in sequential order, the preventive maintenance program (MMS Volume 1) has been developed by the time you are reading this volume. The maintenance work processes (MSS Volume 3), including planning and scheduling, have been implemented. In addition, you should have in place a CMMS / EAM system (MSS Volume 4) with a fully-loaded database. If you were to search the CMMS / EAM system database for all work orders and preventive maintenance tasks for the electricians, you should have a complete duty list.

If you examine the work orders and preventive maintenance tasks at the step level, you would have all of the tasks listed. All you would have to do is add the needs component (knowledge required) and you would have the completed duty–task–need analysis without having to go through the time-consuming interview process. You may want to do a final verification with the subject matter experts, but you will find you have saved a tremendous amount of time compared to a traditional duty–task–needs analysis. In addition, by reviewing who completed the work orders and preventive maintenance tasks, you will also be able to identify subject matter experts for the training program.

Keep in mind that this information is just a sample of what may be found in a task list. Although this task list is somewhat generic, it does help to show the relationship between a duty and a task. For example, if technicians were to demonstrate the knowledge and ability to properly install, inspect, or replace bearings, the first step would be to identify the bearing that was being installed, inspected, or replaced.

To perform this first step or task, determine what information maintenance technicians would need to know. This information forms the needs part of a duty–task–needs analysis (Figure 3-3) on the following page.

This outline shows how to break down a duty into a series of tasks and then to detail the knowledge necessary to perform the task. In the example above, the technicians would need to know and understand the application of each of the various types of bearings in order to install, inspect, or replace a bearing properly.

12.1   Identify common bearing types and their applications.
   12.1.1   Sleeve bearings
      12.1.1.1   Metallic
      12.1.1.2   Non-metallic
   12.1.2   Ball bearings
      12.1.2.1   Single row
      12.1.2.2   Double row
   12.1.3   Roller bearings
      12.1.3.1   Cylindrical
      12.1.3.2   Tapered

Figure 3-3

The needs component of the duty–task–needs analysis forms the outline of the topic that must be presented to the students in the technical training program. In other words, the final output of the duty–task–needs analysis should be a topical outline of a training program that, once developed, should produce maintenance technicians capable of performing the identified duties they may be assigned.

With that being said, unless there are clearly defined expectations of a particular job role, it is virtually impossible to develop a proper training program.

This type of duty–task–needs analysis can be developed for virtually any technical trade that has a detailed job description. What about equipment-specific training programs? For such programs, the process is virtually identical. To begin, one would need to clearly define the duties that a maintenance technician was to perform on a specific piece of equipment. Once the duties were clearly defined, the tasks that needed to be performed to complete the duty would need to be identified. Then, the information needed to perform each task would need to be detailed. This information would make up the outline of material that would need to be developed for the equipment-specific training program.

When there is a duty-task transfer from a maintenance organization to an operational organization, such as an operator involvement, the process is very similar. The duties that are being transferred from maintenance to operations need to be clearly identified. The individual tasks necessary to perform the duties also need to be identified. Then, the information that the operator needs in order to perform each task properly needs to be detailed. This training material may already be developed because the maintenance technicians were likely already performing the work. However, the training material may need to have additional detail added because the operators will not have the technical background. As before, this information makes out the outline of material to be developed for any operator-based activities.

Pitfalls

The pitfalls typical to the duty–task–needs analysis process are listed in Figure 3-4. The biggest pitfall that companies encounter in this process is attempting to take shortcuts. The first major shortcut that companies try is to let an individual subject matter expert make up the duty–task–needs analysis. It is important that a proper cross-section of subject matter experts be selected and interviewed for the analysis. This step, along with selecting a competent facilitator to compile and analyze the results of the analysis, is critical to ensuring that the necessary duties, tasks, and needs are properly identified and documented. If companies attempt to take shortcuts in this process, any training program that is developed will always produce substandard results and little, if any, return on investment.

A second pitfall is conducting the analysis with a preconceived idea of what the results are going to be. This will lead the participants to slant

---

Duty-Task-Needs Analysis Pitfalls

- Shortcutting the duty-task-needs analysis process
- Projecting the results before starting
- Failure to standardize multi-plant activities

Figure 3-4

the results toward their opinion of what the training program should include. Unless the individuals involved can be objective, they should be replaced. If the participants cannot be objective, it is likely the training program that would be developed would be sub-optimized and ineffective.

A third pitfall is the failure to standardize business processes across a corporation. If a training program is developed for multiple plants, it is necessary for all of their business processes and job roles and responsibilities to be the same. As highlighted previously, if this standardization is ignored, a generic training program will result that does not address the real needs of the trainees.

If the guidelines in this chapter are followed, the foundation for developing an effective technical training program will be in place. The next consideration in developing the training program is understand the difference between trainers and trainees.

4

LEARNING AND TRAINING STYLES

People today commonly question the ways in which others people perceive and process information. Whether with frustration, amusement, or acceptance, trainers realize that working with students means dealing with their distinct information handling styles. Each person's different habits and points of view shape their learning styles, which are closely tied to their troubleshooting and problem-solving styles.

In her text *Open Mind,* Dawna Markova identifies six unique patterns of perception. Each one determines how an individual with that pattern absorbs, organizes, remembers, and expresses information. Markova shows how understanding their own perceptual style can help readers increase their self-esteem and relate to others who never seem to understand them.

If they recognize the differences in their students, trainers have an opportunity to adjust their training style and methodology to be more effective. Although it is not possible for trainers to adapt their material to fit each individual student, some adjustment in presentation techniques can make the training more effective. In fact, just by being aware of the strengths and weaknesses of different training and learning styles and methods, trainers can:

- Take an important step toward improved communication with students who don't share their style.
- Build on the strength of their personal training styles.
- Do a better job of designing, developing, and delivering training that accommodates the individual needs of the students.

Background Influences

The individuals in a training setting, both the trainers and the students, inevitably have differences in several areas.

61

Cultural Paradigms

From birth, we vary in our responses to external stimulation. Consider visiting a hospital nursery. As you observe newborns through the large window in the nursery, their behavior may seem to be identical. But as you observe them more closely, you can notice that some infants are more watchful whereas others are more startled by noise. Some are calmer, others more agitated. Yes, from birth, all people vary in their responses to external stimulation.

As individuals grow up in their homes, social environments, and national boundaries, they are influenced by others when it comes to processing and reacting to information. For example, children can have the following experiences:

- Adults may encourage them to touch and manipulate things, to ask questions, and to express their opinions. Or the adults may tell them to be quiet and don't touch anything.
- Children may or may not see neighbors, at work or for pleasure, reading or working with their hands.
- Children may or may not see women or minority group members in positions of authority.
- Children may enjoy school activities and the approval of their classmates and teachers. Or they may have felt confused or excluded. The environment in which an individual was born and raised will play a large role in their cultural paradigms.

Adult Experiences and Learning

Adult students have different levels of formal education and training. An adult may say "I learned my lesson through life experience." What is interesting is that same experience may have led a different person to a different conclusion. A failure may cause one group of individuals to give up, another group to try harder, and a third to stop and think, make changes in their approach, and then try again. As individuals, each person may make a rational, valid choice, or make a choice based on the emotional aftermath of an earlier, perceived failure.

Perception and Skills

Sensory perceptions and physical, behavioral, and cognitive skills differ from person to person. Each individual's bodies and brains, and their interrelated workings, differ. Someone may be left-handed, right-handed, or ambidextrous; another person may be fundamentally a left brain (linear), right brain (holistic), or integrated thinker. Some individuals are morning people, some are afternoon people, and some are night people. Some people score well on tests and some people do not; some individuals who do not do well on tests are still considered smart by their peers. Educational theorists identify the following kinds of intelligence:

- Logical (mathematical) — awareness of groupings, patterns, and sequences
- Linguistic — awareness of shades of differences in word meanings
- Spatial — awareness of shapes or forms and spaces
- Bodily — awareness of bodily control and relationship to physical surroundings
- Musical — awareness of sounds and rhythms
- Interpersonal — awareness of the feelings and intentions of others
- Intrapersonal — awareness of one's own feelings and interests

Individual Differences

Everyone differ in their personalities, feelings, values, viruses, preferences, and expectations. These characteristics affect each individual's way of learning and even their motivation to learn. For example, after a minor training setback, individuals will have different reactions. Some will say, "There must be something wrong with this test." Another group will say, "I know I can do it." Others will say, "I think I should study some more and practice and read the directions again. Then I'll get it." Even others will say: "I know I wouldn't be any good at this; I'll never catch on." Or even: "Not bad for a first try, right?"

Each of these differences in individual students will impact the methods used to present the training and the trainer's method of delivery. Because trainers are going to be with their students for a longer time period, such as an apprentice training program lasting hundreds of hours, they should try to understand how each student is different and how, possibly, to motivate them to learn.

Who's in Charge?

Training styles are related to social and managerial styles. Without a doubt, trainers need to command respect from the students. But commanding respect and commanding the students themselves are two different things.

To begin, who should set the goals and objectives for the program and lessons? There is no right or wrong answer. Trainers may not always be in control of selecting the subject matter or the method of instruction. This decision may have already been made when the training program was designed. For example, if management insists that a group of technical employees acquire certain skills or knowledge, then the instructional designers (possibly the trainers) will work to convert managerial goals into measurable learning objectives.

In doing so, the instructional designers may consider the options of how to present the material to the trainees that would be best suited to their backgrounds and personalities. In this way, the trainees' individuality can be considered in the training design. This style of training is considered trainee-driven. If the course is just to be presented with a generic trainee background (perhaps just considering the basic employee qualifications to hold the job being trained), the training style is considered trainer-driven.

Several factors influence whether the training is trainer-driven or trainee-driven when designing and conducting a training session. These are highlighted in Figure 4-1. They include the following:

Trainer vs. Trainee Driven Training

- Resource Constraints
- Learning Tasks or Materials
- Group and Individual Trainee Needs
- Trainer Philosophy, Knowledge, and Skills

Figure 4-1

a. Resource constraints.

Realistically, there is not always time to allow the students to explore and discover knowledge for themselves. Cramped meeting facilities may crush plans for student-centered activities such as role play. On occasion, a solitary subject matter expert, text, or multimedia resource may be the only available source of information on a topic. When that is true, the question of trainer-driven versus trainee-driven training becomes secondary. Information content (trainer driven) drives the training event.

b. Learning tasks and materials.

Much of the material to be learned — especially tasks such as learning how to repair machinery — requires hands-on experience. Sometimes safety considerations demand tighter trainer control. Such training information is suitable for translation into packets of self-paced instructional material, such as workbooks, videotapes, and computer-based instruction. Because these materials don't take into consideration the trainees' individual needs, they are trainer-driven.

c. Group and individual trainee needs.

How many students will there be? How many trainers? What education, past training, learning tasks, and so forth, do the students have in common? How alike are their learning needs? That is, how similar are they in their preparation for this training, and are all of them to be held to the same standard of evaluation? To what extent are they prepared for independent study? Do they have the necessary level of maturity, training, or experience to managing their own learning? Depending on the answers to these questions, the training can be either trainer-driven or trainee-driven.

Individuals may lack the maturity needed for independent learning. Over time, trainers will meet with typical problem students such as:

- The participants who declare open resistance to the very idea of being trained.
- The "clingers" who are not satisfied with their share of individual attention and so will try to monopolize the trainer.
- The "show offs" who compete with both topic and trainer for the group's attention.

Most adult students are mature enough to recognize and accept their individual responsibility for learning. But are they likely to be accustomed to having training presented to them? And, if they are not training professionals, are they able to find additional learning resources efficiently? With some subjects, it may be appropriate for the trainer to develop alternative materials and activities from which the students may choose. In some cases, it may be worthwhile to create a training course and self-managed learning and to give its graduates access to training materials and meeting space.

d. Trainer philosophy, knowledge, and skills.

What do the session's trainers think about the nature of people? Do they think people are basically industrious, or basically lazy? That they are usually self-propelled or usually need a push? What do the trainers think motivates people — money, status, sense of mastery, fear, or other internal or external factors? How confident are the trainers of their expertise in the training topic? The answer to these questions will determine whether the session is going to be trainer-led or trainee-led.

Trainers need to be careful when making any remarks. For example, if the trainer says "I want you to read section C," the student may think "Who cares what you want? I don't want to do that." A better approach would be for the trainer to tell the students what they will get out of the effort of reading section C. "By using the material in section C., you will find a description and several illustrations of the type of maintenance activities you working on next week. This material will make the task easier." Most students will quickly read section C.

David Kolb, the author of *Experiential Learning: Experience as a Source of Learning and Development*, theorizes that learners are oriented towards four learning modes. They are:

• Concrete experience

• Reflective observation

• Abstract conceptualization

• Active experimentation

Concrete experience emphasizes feeling as opposed to thinking. People with this learning orientation take an artistic approach. They are intuitive, open-minded, and do well in the absence of structure. Trainers of people with this learning profile should approach the training with the view to be a motivator.

Reflective observation involves understanding the meaning of ideas and situations by carefully observing and impartially describing them. People with this learning orientation can see the implications of different approaches and are good at understanding different points of view. Trainers of people with this learning profile should approach the training as if they are the experts.

Abstract conceptualization concentrates on thinking as opposed to feeling. People with this orientation like to take a scientific, systematic approach. They like working with symbols and analyzing information to formulate general theories. Trainers of people with this learning profile should approach the training as coaches, providing guided practice and feedback.

Active experimentation focuses on actively influencing people in changing situations and emphasizing practical applications. People with this orientation like to get things done. Instructors of these people should stay out of the way, providing them maximum opportunities for discovering themselves.

Although it is tempting to try to tailor training to the learning styles of the learners, it is not very practical. Training in the workplace must be designed to meet the requirements of the performance task rather than to accommodate any particular learning style.

Awareness of style differences, both the trainers' and the students', is useful in several ways. First, it captures the differences among trainees. Second, it should encourage trainers to use a wider variety of teaching techniques. Third, it helps students to understand their own styles and preferences better. Finally, it enables trainers and students alike, to be more accepting of the differences among people. In this manner, new means of communication are established in the diverse audiences that most trainers encounter in today's workplace. Whatever their own styles are, today's trainers and developers have much to think about and do when considering the continuing research and application of training and learning styles.

The Trainers

We also need to realize that there are different types of trainers and training styles (see Figure 4-2). Some of these styles can be quite negative.

- The Professor — This individual dispenses knowledge and approval, but does not expect the class to value learning for its own merit.
- The Comedian — They keep a class laughing, but are not concerned about what the class learns.
- The Projectionist — They are audiovisual, computer wizards, but apart from getting the tape, film, or computer program running, they do not feel obliged to offer the students help.
- The Inspirer — They get classes all worked up, but offer no substance to work on.
- The Drill Instructor — They think the class members are dull and lazy and concentrate on repetition, repetition, and more repetition.

> # Trainer Profiles
> - The Professor
> - The Comedian
> - The Projectionist
> - The Inspirer
> - The Drill Instructor

Figure 4-2

Each trainer stereotype derives from an unbalanced overconcentration on one of the following desirable trainer traits:
- Knowledge of subject
- Sense of humor
- Technical skill
- Awareness of others
- Willingness to lead when necessary

How can instructors discover their training style? Their strengths and weaknesses? Comments from trainees and, of course, their evaluations can be a starting point. If trainers are surprised by the comments on an evaluation, they are seeing a personality trait that is in their own personal blind spot.

Trainers have four aspects to their training skills. The first is that part of their personality that is known to themselves and others. The second is that part of their personality that is known to themselves, but not known to others. The third is that aspect of their personality known to others, but not known to themselves. Finally, the fourth covers those personality aspects not known either to one's conscious self or to others. Trainers need to evaluate themselves constantly — both from a personality perspective and a technique perspective. Comments from training participants and from peers are useful tools for finding and eliminating "blind spots" in their personalities.

Trainer Styles

Trainer styles can be classified in at least four basic categories.

- Listeners — They create an effective learning environment that encourages learners to express personal needs.
- Directors — They create a perceptual learning environment that provides learners different perspectives.
- Interpreters — They create a symbolic learning environment that encourages learners to memorize and understand the terms and rules.
- Coaches — They create a behavioral learning environment that allows learners to experiment and evaluate their own progress.

Again, if trainers are serious about developing and maintaining their training skills, they will be attentive to feedback from their peers and their students. Once trainers recognize their strengths and weaknesses, they can do much to improve their skills. Discerning the types of students they have in a training session and addressing the students' differences (in light of the instructor's style and profile) will do much to enhance the learning experience for the students. And a secondary benefit —the training activities will be more rewarding for the trainer.

5

Preparing Instructional Objectives

Following the process outlined in Chapter 2, the next step in the training program development is the preparation of the instructional objectives. What material do the trainers need to teach and how will they teach it? Before selecting a book or a video on a particular topic, trainers must understand the actual objectives of the training.

Instructional objectives are critical to developing focused, cost-effective, technical training programs.

Considerations for Instructional Objectives

- They reflect organizational needs and goals.
- They reflect the training population's needs.
- They allow test items to be written for each objective.
- They allow test items to match a related objective.
- They allow training strategies to reflect resource constraints, but honor the training population needs.
- They insure logical training sequences can be determined.

Training is effective to the degree that it succeeds in changing the trainees, usually improving their knowledge and skills — and not in undesirable directions. Training that doesn't change the trainees positively has had no effect and is a waste of resources. If the training changes the students in undesirable directions (perhaps by influencing their opinions about training in a negative way), it is worse than ineffective. Instead, it wastes time and money, and it sets the organization back. Training is successful or effective only to the degree that it meets organizational needs and goals.

Once the decision has been made to train someone, several kinds of activities are required if the training is to be deemed successful. For one thing, the responsible individuals (usually senior management) must be assured that there is a need for the training. The trainers must make certain that the students don't already know the material and that training is the best means for bringing about the desired change. Another activity requires the trainers to specify clearly the outcomes they intend the training to accomplish. The trainers must then select and arrange training experiences for the students in accordance with the agreed-upon principles. Finally, they must evaluate student performance in accordance to the objectives originally selected. In other words, trainers first determine the gap between the students' current state, then create and administer the means of getting them to the new, desired state. Then the trainers must evaluate whether of not the goals have been achieved.

The steps for accomplishing successful training programs arrange themselves into these four main phases:

1. Analysis
2. Design/ development
3. Implementation
4. Evaluation/ improvement

A number of procedures and techniques are available to help complete these four phases. For example, the analysis phase should answer questions such as these:

• Is there a real skill deficiency?
• Is training the solution?
• If so, how will the training provide the solution?

After all, training is only one of several possible solutions to problems of human performance. Unless a suitable analysis is performed before the training is developed, it is quite possible to construct a great training course that doesn't help anybody at all. It is possible to construct a course that no one needs, either because it is unrelated to solving the skill deficiency that gave rise to the need for a course or because it teaches things the trainees already know. Techniques such as the duty–task–needs analysis that was considered in Chapter 3 will help avoid wasting resources.

If the duty–task–needs analysis reveals that training is needed, the instructional objectives are drafted that describe the important outcomes intended to be accomplished by the training. In other words, instructional objectives are drafted that answer the question "What is worth teaching?" Assessments, by which the success of the instruction can be determined, are then drafted.

This process can be more complicated if a multi-plant or corporate training program is being developed. As pointed out in Chapter 3, input from all of the plants involved must be gathered. This input must then be consolidated. Once it is consolidated, the instructional objectives will then need to be developed and the assessments drafted. However, as simple as the process may appear, the same issues highlighted in Chapter 3 will need to be re-considered when dealing with multiple plants. These include:

• Difference in the duty-task-needs analysis
• Differences in roles and responsibilities
• Differences in work processes

Only after the preceding steps have been completed is the actual training drafted, tested, revised, and then utilized. And only after the needs analysis phase is complete, or is near completion, are the instructional objectives drafted. This is an important point because trainers need to hear that the first step is to write objectives — before instruction is designed.

What Is an Instructional Objective?

An instructional objective is a collection of words or pictures and diagrams intended to let others know what the trainers intend the students to achieve.

The instructional objective is related to intended outcomes of the training, rather than the processes for achieving those outcomes. It is specific and measurable, rather than broad and intangible. It is concerned with the trainees, not the trainers.

The goal of this chapter is to enable the reader to develop well-stated objectives. Given any objective in a technical subject area with which you're familiar, you should be able to identify correctly the performance, the conditions, and the criteria of acceptable performance when those characteristics are present.

Outcomes vs. Processes

The objective is related to an intended outcome of training rather than the process of training. For example, lecturing is something an instructor does to help the students learn; it is part of the process of instruction. The lecture is not the purpose of the instruction. The purpose of the instruction is to facilitate learning. So when teachers teach, they do it because they hope the students will learn. Therefore, statements such as the following are descriptions of instructional process rather than of the intended results:

- Provide a lecture series on hydraulics.
- Be able to perform well in a role-play situation.
- This course provides extensive practice sessions.

Because recognizing the difference between process statements and outcome statements is critical to the effective use of objectives, it's necessary for trainers to develop the ability to spot the difference.

Think of the training as being like a vehicle that takes trainees from one place to another. The question to be answered by an objective is "What are the students expected to be like when they arrive at their destination?"

It might help to think of the difference between statements describing the process of building a house and those describing the characteristics or outcomes of the completed house. For example, here are some process statements about the construction process:

- The foundation is laid before the walls go up.
- Walls are to be constructed of Pine 2 x 6s.
- Scaffolding will be used when installing the roof.

In contrast, the following statements describe the characteristics or outcomes of a completed house:

- The house contains three fireplaces.
- The front of the house faces south.
- All windows are constructed of double pane glass.

The outcomes are the results we get from the processes, but they are not the processes.

Specific vs. General

Another characteristic of an objective is that it is specific, rather than general, broad, or hazy. If objectives are hazy, they don't do the trainers or the trainees any good, and the trainers might as well not bother with them. Instructional objectives need to be specific so they will help the trainers make good instructional decisions.

Specific instructional objectives are precise. After reading them, you can immediately determine whether or not trainers and trainees have met the objective. If the objective has not been met, the specific statement indicates what must be done to meet it. On the other hand, general or abstract statements leave trainers in the dark. The instructional objective must be reworded until it says exactly what is expected.

Measurable vs. Immeasurable

An instructional objective is considered measurable when it describes a tangible outcome. For example, objectives that describe intended outcomes that can be seen or heard or are measurable.

Consider the objective "To be able to tap a 3/8" UNC bolt hole." This objective is measurable because we could see the knot tying behavior and, therefore, assess whether it meets our expectations.

On the other hand, a statement that says "To be able to internalize a growing awareness of confidence," is not measurable. It can't even be called an objective. What would you measure? What would you watch a student do to decide whether or not internalizing had occurred to your satisfaction? The statement doesn't say.

Trainees vs. Trainers

Instructional objectives describe the trainee's performance rather than the trainer's performance. Objectives that describe the trainers' performance are called administrative objectives, not instructional objectives. Trainers help students accomplish the instructional objectives. Many trainers cannot distinguish between statements about trainer activities and trainee performance. If an objective is going to be useful, it needs to contain specific and measurable trainee outcomes.

Why Care about Objectives?

Training is only successful to the degree that it succeeds in changing trainees, allowing them to achieve a desired behavior and modifying or eliminating an undesirable behavior. If the training doesn't change the trainees in desired ways, it isn't any good, regardless of how elegant the lectures are or how complicated the presentation mechanisms.

If training is to accomplish desired outcomes, it is imperative that those designing the training, as well as those actually doing the training, have a clear picture of those desired behavioral changes. Because instructional objectives are tools for describing intended outcomes, they provide a key component for making training successful and are useful in many ways.

Materials/Procedure Selection

When clearly defined instructional objectives are lacking, there is no sound basis for the selection of training materials and procedures. After all, machinists or electricians don't select tools until they know what they're intending to accomplish. Too often, however, one hears trainers arguing the relative merits of books versus lectures, computers versus video, self-pacing versus group pacing, without ever specifying just what results they expect these things to achieve. Trainers simply run into in a fault of their own making, unless they know what they want their trainees to accomplish as a result of the instruction.

Instructor Ingenuity

Once the important outcomes of the training have been understood and clearly stated, it is possible to say to the trainers, "Here are the objectives you are expected to achieve. Now go use your best experience and ingenuity to achieve them." Thus, the existence of the objectives frees the trainers to be creative and flexible. With the instructional objectives in place, it is no longer necessary to expect all trainers to be doing the same thing at the same time during a lesson.

Consistent Results

Instructional objectives provide the basis for achieving consistent training results. With the instructional goal post clearly visible, it is possible to provide enough training and practice so that all trainees learn to perform at least as well as the instructional objectives require. Some trainees will learn more or reach a higher performance level than the objectives require, but every trainee can and should be expected to accomplish at least each objective.

With instructional objectives, it is possible to achieve desired results without requiring consistency in the process the trainers will use for getting those results.

Measurable Results

How many courses have you taken in which the tests had little or nothing to do with the substance of the instruction? Unless instructional objectives are clearly and firmly fixed in the minds of both instructors and students, tests are likely to be misleading at best; at worst, they will be irrelevant, unfair, or uninformative. Without clear instructional objectives, it simply isn't possible to decide which measuring instrument will tell you what you want to know.

Clearly-stated instructional objectives provide a sound basis for selecting the means by which to find out whether they have been achieved. Suppose part of an objective states "to be able to disassemble, repair, and reassemble a gear case." How could trainers find out whether the trainees can actually do what they are supposed to learn to do? How about a multiple choice test? After all, they are easy to grade, and the trainer could even claim to be using an objective test. Perhaps true–false questions would be better? Trainers can have a lot of fun dreaming up wrong answers. How about an essay test? Trainees should be able to describe how to take a gear case apart, and how to put it back together, shouldn't they?

No doubt, it can be seen right away that the only way to test whether someone can disassemble, repair, and reassemble a gear case is to say "Let me see you disassemble, repair, and reassemble a gear case." How would trainers know that? The objective would clearly state the intended outcome of the instruction. With clear instructional objectives, trainers don't have to be experts in test construction to select and create measuring instruments that will tell you whether your objectives have been accomplished.

Goals for Trainees

Clearly-defined instructional objectives can also be used to provide trainees with the means to organize their own time and efforts towards the accomplishment of the objectives. When the instructional objectives have been clarified and revealed to the students, it is no longer necessary for them to guess what the instructors might have in mind for them to accomplish.

As you know, many trainees are required to spend considerable time and effort learning the peculiarities of their trainers. Often those trainers fail or refuse to let their students in on the secret of what they're expected to learn. Unfortunately, such knowledge can be useful in helping students breeze through courses with little more than a bag full of tricks designed to gain favor with the trainers. Clear instruction objectives in the hands of the students eliminate the need for such time-wasting and anxiety-producing activities.

Instructional Efficiency

It has frequently been observed that when good objectives have been developed, existing instruction often can be dramatically shortened. In fact, training can sometimes be eliminated altogether when the objectives reveal that inadequate job performance is due to factors other than lack of knowledge or skill.

This result is affected by comparing information about what trainees need to be able to do vs. information about what they already know how to do. When there is no difference between what they "should be able to do" and what they "can already do," more training will not help. In these cases, the source of inadequate performance must be found somewhere else. There are many reasons why people don't perform the way they are supposed to, such as unclear roles and responsibilities, unclear job instructions, absence of tools, space, or authority to perform as expected.

Instructional objectives are also helpful for organizations responding to the pressures of downsizing that create the need to do more with less. With such pressures operating, it is critical for employees to become competent as quickly as possible. At the same time, it is important that the employees not be removed from their job sites any longer than absolutely necessary to attend training. Instructional objectives not only allow the training to be streamlined to the needs of the individual trainee, they often allow the training to be delivered a module at a time, at the most convenient locations, and during short periods that do not disrupt the flow of work.

In summary, instructional objectives are useful for providing a basis for selecting training materials and procedures, room for trainer's creativity and ingenuity, measurable training results, tools for guiding trainee efforts, and a method for evaluating instructional efficiency.

There are additional advantages, not the least of which is that the act of drafting objectives causes one to think seriously and deeply about what is worth teaching. When instructional objectives are drafted for courses already in existence, they can serve to spotlight opportunities for instructional improvement.

Where Objectives Originate

Instructional objectives come from many sources. Some of these sources are rational, systematic, and useful. Other resources are egocentric, disorganized, and amazingly haphazard. Systematic processes to develop the instructional objectives lead to the ones truly worth accomplishing. The "I know what's best for students" approach, on the other hand, often leads to objectives that describe outcomes of little value to the trainee. This is because the "I know best" decisions can often be totally disconnected from any real need for training. Such questionable decisions can be derived from prior experience, which may be out of date, from biases inspired by the chapters that happen to be included in a textbook, from the trainer's preferences about what they like to teach, or from inertia, such as "I've always taught it this way."

In the training program detailed at the start of Chapter 3, the instructors were actually encouraged to limit their tenure in the program to a 3–5 year window. This approach always kept "fresh" trainers in the program and prevented many of the problems mentioned in the previous paragraph.

When derived from any of these non-systematic methods, the resulting training can prove totally useless to the trainee, regardless of the importance of the subject matter to the trainer. Unfortunately, people embedded in the middle of an educational system can easily lose sight of the fact that good objectives are ultimately derived from the real world. This is another way of saying that the purpose of training is to help someone learn to do something of value to someone other than the trainer. Sometimes, trainers can get so engrossed in training points that they forget that the training should be designed to get beyond the talk and move the trainees into job performance.

Properly derived instructional objectives, genuine objectives, are ultimately about job performance. They describe the desired results of the training, rather than the action of training. They provide descriptions of training outcomes, thereby, allowing the trainers to identify the parts of the instructional process that will truly be relevant to teaching those desirable outcomes.

Where to Identify Instructional Objectives

Instructional objectives can be developed from the duty–task–needs analysis, as discussed in Chapter 3. All jobs consist of a collection of tasks that must be performed. This collection of tasks makes up the list of the things that people should do when carrying out their jobs. If the job responsibility is new, or if there is some question about whether the current tasks are being performed appropriately, a higher level of analysis will be required. In other words, first it should be determined what the job should consist of (the duties). Then the tasks that will be required to complete the job (or duty) properly should be listed.

Once these tasks have been identified, the next step is to diagram each task, describing the steps and key information that will be required to complete each task. This process, called tasks detailing, reveals the components of the task by describing what a competent person, such as an SME, must know and do to perform the task properly. The diagram states the reason for starting to perform the task, it describes the steps that follow, and lists the necessary information.

Consider the example from Chapter 3 of the duty–task–needs analysis.

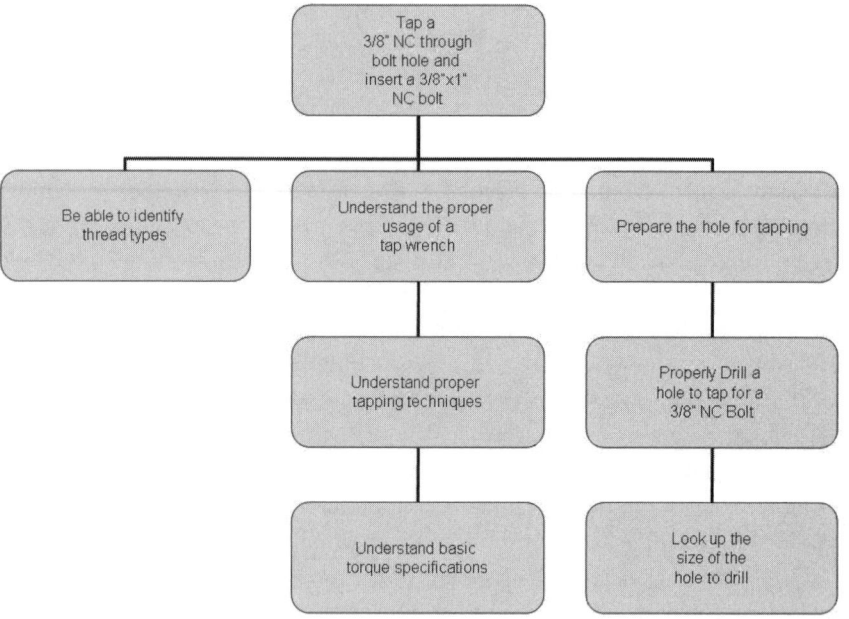

Figure 5-1

The duty:

12. Demonstrate knowledge and ability to properly install, inspect, or replace bearings.

Figure 5-2 shows what a sample list of tasks that might follow.

12.1	*Identify common bearing types and their applications.*
	12.1.1 Sleeve bearings
	12.1.2 Metallic
	12.1.3 Non-metallic
	12.1.4 Ball bearings
	12.1.5 Single row
	12.1.6 Double row
	12.1.7 Roller bearings
	12.1.8 Cylindrical
	12.1.9 Tapered
12.2	*Demonstrate ability to properly mount and dismount bearings.*
12.3	*Identify the proper bearings seals for specific applications.*
12.4	*Demonstrate ability to properly lubricate various bearing types.*
12.5	*Identify various bearing wear patterns.*
12.6	*Demonstrate ability to find bearings in appropriate catalogs.*

Figure 5-2

If the duty is 12. Demonstrate knowledge and ability to properly install, inspect, or replace bearings, then 12.1–12.6 are the tasks that the trainees will need to be able to perform. In turn, the need information for task 12.1 is the ability to identify the bearings listed in 12.1.1–12.1.9. The trainees would then need to know the information in 12.2–12.6 for each of the types of bearings listed in 12.1.1–12.1.9. With these requirements detailed, instructional objectives can be developed to specify the course requirements.

With a task analysis in hand, it is possible to answer the question, "What would someone have to know or be able to do before being ready to complete this entire task?" The list of required skills can be systematically derived from a listing of what competent technicians (SME) actually do.

With this list of required skills having been derived from the duty–task–needs analysis, the next step is to draft objectives describing the level of skill that someone would need to perform the various tasks. For example, would a new employee be expected to have the skill? What about an apprentice, or a journeyman technician?

The objectives describing the skills needed for the performance of all the job-related duties-tasks provide the basis for the development of a training module or a coaching session. The beauty of this procedure is that the trainers who have developed objectives in this manner are able to prove that what they are teaching is relevant to the fulfillment of an important job need.

The next step is to draw a skill hierarchy that shows the prerequisite relationships between the objectives. A hierarchy looks a great deal like an organizational chart and shows which skills need to be mastered before another can profitably be practiced. For example, if trainees need to drill and tap a 3/8" bolt hole, there are some prerequisites before they can accomplish that task satisfactorily. They will need to understand thread types, how to drill a hole, and the size of hole that is required. They will also need to understand how to turn the tap, yet avoid breaking it due to material jams or from applying excessive pressure. This hierarchy is illustrated in Figure 5-2.

To this point, the focus has been on what anyone would have to do to be able to perform competently in a particular job assignment. With the objectives and hierarchy in hand, it is now possible to derive an efficient course outline for each trainee by comparing the objective with what a given trainee can already do. If a given trainee can already complete one or more objectives (or demonstrate proficiency), these objectives are deleted from that particular trainee's curriculum.

Trainers who are working in business and industry today must understand that they will be expected to develop the instructional objectives systematically, following the steps just described.

Creating Objectives

How can a trainer create effective objectives? What characteristics will make an objective able to communicate clearly to the trainees and be useful to them? Several schemes can be used in stating objectives, but the following format is one that will work in a technical training environment.

Three characteristics — performance, conditions, and criteria — are especially important to include in an instructional objective in order to communicate the intent of the objective.

- Performance. What should the trainee be able to do or produce to be considered competent?

- Conditions. Under what conditions should the trainee to be able to complete the objective?
- Criteria. How well must it be done — what measurable outcome must the trainees produce in order to be considered competent?

Sometimes no special conditions are included. Sometimes it is impractical or useless to include criteria (because the criteria required are obvious). But the more the trainer says about their desired intent, the better they are communicating.

Performance

To be usable, an instructional objective must state what performance trainees must achieve to demonstrate their mastery of the objective. This statement is easy to do when the main intent of the instructional objective is visible or audible. For example, (referring back to the duty-task-needs analysis in Chapter 3) if the objective calls for the trainees to perform a specific repair, demonstrate specific communication skills, or draft a failure analysis report that has specific characteristics, the objective will state these visible and audible requirements.

If a statement does not include a visible or audible performance, it is not yet an objective. Many statements, however, only pretend to describe a performance. In such cases, it is highly unlikely that two or more people could agree on what the statement means because it is open to too many interpretations. As an objective, such a statement is useless because it doesn't clearly communicate the intent.

For example, the statement "Demonstrate good repair skills while rebuilding a gear case" is vague. What makes one repair good and another weak? A better objective would be "Demonstrate the following skills while rebuilding a gear case" and then list specific skills such as applying proper torque values to all fasteners, properly installing the gasket, properly aligning the bearings to 0.003, properly adjusting the gear backlash to 0.005, and so forth.

Suppose the objective was stated as "Develop the proper skills to rebuild a gear case." The word "develop" suggests that the statement is referring to the process by which someone might come to have an understanding of how to rebuild the gear case. By contrast, "Demonstrate the proper skills to rebuild a gear case," including:

- Applying proper torque values to all fasteners
- Installing the gasket properly

- Properly aligning the bearings to 0.003
- Properly adjusting the gear backlash to 0.005 is a more specific objective.

All the components of this objective have visible and measureable performance criteria for auditing the outcome of the trainees' efforts.

Objectives are about intended outcomes. Therefore, the defects in the following statements should be obvious:

- To develop,
- To solve,
  To discriminate,
- To identify

These objectives are all partial or unclear. Although these words can be used in objectives, they will need further clarification for them to have visible or measurable results. The reason is they don't tell us what trainees must do to demonstrate they can complete the objective. The important performances the trainers intend to have the trainees develop are unclear and the measurement of the results of the training is now subjective.

How can this problem be solved? Whenever the instructional objective's main intent is unclear (usually when the instructional objective consists of a verb only), an objective clarifier should be added. The clarifier will more directly specify the performance. An objective clarifier will set certain criteria to insure the trainee performance is achieving the desired level of performance. An objective clarifier is simple and direct, and is always something that every trainee already knows how to do. For example, if the instructional objective is about the ability to determine the difference between a good and defective bearing by a visual observation, the objective might state "After the completion of this training unit, trainees should be able to identify defective bearings."

Objective clarifiers that are added should be simple, direct, and specific. They should be based on directions that trainees would already know how to do. Common examples of objective clarifiers are circle, underline, point to, write, or say. Keep in mind that instructional objectives are not about the clarifiers themselves. They are about main intents. It's the main intent that trainees are supposed to learn. If they have to learn an objective clarifier, that clarifier should not be used.

To summarize:

- An instructional objective describes an intended outcome of the training, rather than the procedures for accomplishing those outcomes.
- An instructional objective always states a performance that describes what the trainee will be doing when demonstrating mastery of the objective.
- When the main intent of instructional objective is vague, an objective clarifier is added, through which the main intent can be detected.
- Objective clarifiers are always the simplest, most direct behaviors possible, and they are always something that every trainee already knows how to do well.

Conditions

In some cases, specifying the performance may not be enough to prevent a serious misunderstanding of the instructional objective. To avoid surprises when working with instructional objectives, trainers must state the main intent of the instructional objective and describe the main condition under which the performance is to occur. An instructional objective that says "Hammer a nail" is different from one that says, "Using a brick, hammer a nail." You might assume that the first objective "Hammer a nail," means hammering with a hammer.

This assumption would be logical in the absence of any other information. But think how dismayed the trainees would be if they practice hammering with their hammers, but when it came time to demonstrate their skill, they are asked to hammer with bricks (or crescent wrenches or pieces of pipe). They would undoubtedly feel betrayed, tricked, and deceived — and they would be right.

Miscommunications can be avoided by adding relevant conditions to the instructional objective. Avoid confusion by describing the conditions that have a significant impact on the performance of the instructional objective. When trainees are told what they will and will not have to work with when performing, and are also told of special circumstances in which the performance must occur, they will be in a greater position to understand the goal of the instructional objective.

How Much Detail?

How detailed should trainers be in their descriptions? They should include enough detail to describe each of the conditions that would be needed to allow the specified performance to happen. The trainees should be aware of any conditions that would make a significant difference to their performance. How should the conditions be described? Should every instructional objective have conditions? Not necessarily. There should be enough detail that the trainers and the trainees both are clear on the objective. Consider the gear case example again:

"Demonstrate the proper skills to rebuild a gear case," including:

1. Applying proper torque values to all fasteners
2. Installing the gasket properly
3. Properly aligning the bearings to 0.003
4. Properly adjusting the gear backlash to 0.005

The third and fourth items have measureable parameters that will be clear to the trainers and the trainees. However, what measurement device is to be used? A micrometer? A dial indicator? A feeler gauge? All will have different tolerances and will give different results. The first skill listed also may need a condition specified —the size of the torque wrench that will be used. Should it be a 1/4" drive? A 3/8" drive? A 1/2" drive? They will all torque to a certain specification such as installing a fastener to 10ft-lbs of torque. But what if more is required? Then the trainees could quickly exceed the measurement capacity of the 1/4" torque wrench and could even exceed the measurement ability of the 3/8" drive.

To make the first skill more precise, the objective could read:

1. Using a 1/2" drive torque wrench, apply the proper torque values to all fasteners.

When determining whether or not to use conditions, consider what is expected from the trainees. If just stating the desired performance and the degree of excellence makes clear what the trainers desire, then don't add conditions. How can trainers tell whether the conditions are defined clearly enough? They can give their draft objectives to a couple of trainees and ask them what they think they would need to do in order to demonstrate their mastery of the objective. If their description matches what the trainers have in mind, then the trainers have done well. If their description doesn't work, then an adjustment or additional description is in order. But remember the ironclad rule of instructional objective writing: if there is a disagreement about the meaning, don't argue about it. Just fix the objective.

Criteria

Once trainers describe what they want the trainees to be able to do (their performance), and the circumstances in which the trainer wants them to do it (the conditions), the trainers will have given the trainees far more information about their intention than trainees are accustomed to receiving. There is something else trainers can do, however, to increase their communication about their objectives. They can add the criteria of acceptable performance. This information will tell the trainees how well they will have to perform in order to be considered competent. By adding this information about the measurement of the instructional objective, the trainer will strengthen the usefulness of the objectives. The following list names several advantages of adding criteria:

- The trainers will have standards against which to test the success of the instructions.
- The trainees will know how to tell when they have met or exceeded the performance expectations.
- The trainers will have the basis for proving that their trainees can do what the trainers set out to teach them.

These are powerful benefits in the quest for improved performance. What the trainers must do is complete their instructional objectives by adding information that describes the criteria of success. If the objective isn't measurable, it isn't instructional. Here the focus is measurement. How can trainers add criteria of acceptable performance to objectives?

Trainers should not specify either a minimum or barely tolerable criteria. Instead, they should look for ways to describe the desired or appropriate criteria. Sometimes that means a low performance level is acceptable, and considerable error can be tolerated. Other times, only perfect performance is acceptable —no errors can be tolerated. For example, it might be acceptable at times for a maintenance technician to skip a step on a preventive maintenance checklist (especially if a repair has been performed recently). However, airplane pilots are expected to check every task on the pre-flight checklist. Every time!

Time limits are often used as criteria for acceptable performance — given performance must occur within a certain length of time. Such a time limit is often implied when a trainer tells the trainees how long an examination will be. If the speed of performance is important, however, it is better to be explicit about it with the trainees. Then they will not have to guess what the trainers have in mind for them to do. When time is important, it is only fair to communicate that information to the trainees.

Accuracy is another criteria of acceptable performance. In some cases, accuracy may be combined with speed as dual criteria.

Sample objectives which include accuracy as a criteria include:

- Read the measurement on a micrometer to 0.001" of accuracy.
- Solve the math problem to the nearest whole number.
- Weigh the materials to the nearest gram.
- Write the answer to at least three significant figures.
- Make no more than two incorrect entries for every 10 pages of work orders performed.
- Listening carefully enough that no more than one request for repeated information is needed for each set of verbal job instructions.
- Finish all machined surfaces to a 64 smoothness tolerance.

The word descriptors or measurements will communicate how well the trainees must perform before the trainers will agree that the trainees have achieved the objective.

In some cases, the speed or accuracy of the performance are not the measure of acceptable performance. Instead quality is the measure. For example, the objective might be to "machine bar stock to a dimension of 1/2" x 2" x12" plus or minus 1/32" in length. The quality concern is the 12" dimension being within 1/32". How fast the trainee was to perform the work was not the issue. Furthermore, the accuracy of the other dimensions was not specified because they are usually controlled by the manufacturer of the bar stock.

Selecting Criteria

How do trainers know whether criteria should require that a performance be completed in 10 minutes or 20 minutes? Or that it should demand a hydraulic cylinder extend with a speed of 10 ft/min or 5 ft/min? Or that the objective should call for completing 5 PMs per day instead of 2 PMs? There are two sources from which these criteria may be derived.

The first is job requirements. Those who derive objectives from real-world needs will observe and interview competent or exemplary performers, then describe what the performers do and how well they do it. Next, they will use this information as the basis for deriving instructional objectives. At this point, they will further identify the criteria that should be attached.

Remember that not all objectives require perfect performance in order to be achieved. Otherwise, the criteria may not be realistic. Perfection costs money — the tighter the criteria, the costlier the training. Consider smoothness. The smoother that you want a machined surface, the more it will cost. Any time you see an instructional objective that calls for perfection, question it. Make the objective writer defend that criteria as realistic.

Sometimes the criteria are set at an entry level. In these instances, the criteria reflect what employers require for entry-level skills rather than the skills they expect employees to have after extensive experience and practice. When managers say "I can't hire electricians who can run conduit," it is important to find out what they mean when they says "run conduit" — straight, level, other? Then set the criteria to insure employees will be able to meet the requirements.

A second source is personal experience. This is another guide to criteria that comes from the personal experience and wisdom of expert technicians. People who are actually performing the skills described by the instructional objectives in real-world settings are likely to have good insights into the performance quality needed to do the job well (especially maintenance planners and supervisors — See MMS Volume 3). Thus, if trainers are drafting instructional objectives — and at the same time are performing the skills they are writing about — their judgments about appropriate criteria would be valuable. However, if they have never performed the task that the objectives describe, they should not rely on their judgments. They should go to competent performers (SMEs) and derive the criteria from their performances.

Trainers have several ways in which they can indicate criteria without explicitly describing them. The intended criteria may have been made explicit in a related document. The objective can then include added words to tell the trainees where to find the criteria. For example, the objective might be to adjust the gear backlash to the manufacturer's recommendations in the equipment manual, which is stored in File Drawer #32 in the maintenance office.

A checklist may already exist consisting of several steps that make up the desired performance. In this case, the trainers might point to that checklist as part of the criteria, rather than explicitly listing the steps themselves. For example, a slightly different objective than the one in the previous paragraph might read, "Rebuild the gear case by following the checklist in the manufacturer's manual, which is stored in File Drawer

#32 in the maintenance office."

On other occasions, the trainers might find it appropriate to point to a competent video performance shown on a piece of film, DVD, CD, or other, in effect saying to perform the task as shown in the media. The video demonstration might be useful when a required performance involves complex movements that are difficult to describe. This method should not be a license for the trainers to use only dynamic media, without describing the key characteristics of the desired performance in the objective itself. Such a practice would be as uninformative as other false criteria such as "to the satisfaction of the trainer." Refer to film, DVDs, and other samples only if they help in making the desired criteria clear to all.

When preparing instructional objectives, trainers should continue to modify them until these questions are answered:

- What do I want the students to be able to do?
- What are the important conditions or constraints under which I want them to perform?
- How well must the students perform for me to be satisfied?

Pitfalls to Developing Instructional Objectives

False Performance

This point was made earlier, but deserves some repetition. All too often statements are mistakenly called instructional objectives because they have the appearances of objectives, yet they contain no performance standards. Therefore, they are not objectives at all. When statements without performance standards are thought of as instructional objectives, they lead to a large amount of confusion. Trainers and trainees are likely to argue about which instructional procedure is suitable for accomplishing the vaguely-stated intent; they are frustrated when the statement offers no firm guidelines. In addition, trainers cannot agree on methods for assessing achievement of the intent and may complain that all objectives are useless. The trainers are then at a loss in understanding why the trainees are themselves at a loss in understanding what they are expected to be able to do. Little wonder, as broad statements provide few clues to action.

When interpreting or drafting an instructional objective, trainers must first look for the desired performance to be achieved. Once it is identified, highlight it. If you can not highlight a desired performance that will be achieved, the proposed instructional objective is not an objective.

Revise it or discard it. Consider the gear case example, once again. By itself, "Rebuild a gear case" does not provide sufficient information for it be considered an instructional objective, for all the reasons we have considered thus far through this chapter. There is no way to measure the performance, since no clear performance is specified. How long should the rebuilt gear case last? What are the tolerances? What tools should be used?

False Givens

Another common error is the inclusion of false givens. These are words or phrases that follow the word "given" in an instructional objective. In this case, they describe something other than the specific conditions the trainees must have (or something they will be denied) when demonstrating the achievement of the objective. Typically the words describe something about the instruction itself. As indicated earlier, an instructional objective is useful to the degree that it communicates an intended outcome. If the trainers allow it to describe an instructional procedure, then the trainers and the trainees are restricted from using their best wisdom and experience to accomplish that outcome. Trainers need to assure that the conditions described in the objectives tell something about the situation in which they expect the trainee to demonstrate competence.

Suppose the objective reads, "Given the information provided in class, demonstrate the proper skills to rebuild a gear case." This objective indicates that the trainees will be provided sufficient information in the course itself to demonstrate the proper skills. The "given" portion of the objective is misleading — you should assume that the trainees will have sufficient information by the end of the class. What's not provided in this objective is the description of what would be considered proper skills. Again, consider the more accurate instructional objective: "Demonstrate the proper skills to rebuild a gear case," including

1. Applying proper torque values to all fasteners
2. Installing the gasket properly
3. Properly aligning the bearings to 0.003
4. Properly adjusting the gear backlash to 0.005

Teaching Procedures

Related to false givens is the error of writing an objective to describe a teaching point, a practice exercise, or some other aspect of classroom activity. There are two practical considerations. First, if the trainer describes all instructional activities or teaching points and calls them instructional objectives, the objectives will drown in verbiage. Second, the main function of the instructional objective is to help course planners decide on instructional content and procedures. If the objective describes a teaching procedure, it will fail to perform its primary purpose, because it will be describing instructional practice rather than important instructional outcomes.

Trainers can avoid this problem by asking themselves why they want the trainees to be able to do what they have described in each objective. If the answer is because that is one of the things they need to be able to do when they have finished the training, the objective can probably stand. However, if the answer is, so they can complete a task, but what it takes to do the task is not stated in the instructional objective, then it is likely a teaching procedure, not an instructional objective. If we consider the example of rebuilding a gear case, a bad objective would be:

"Using a torque wrench, demonstrate how to rebuild a gear case."

This is a teaching procedure — telling the trainers and trainees what tool to use. It is actually part of fulfilling the instructional objective of "Demonstrate the proper skills to rebuild a gear case," introduced above.

If the objective falls into this pitfall, it should be modified to describe the desired outcome. If trainers always write instructional objectives about the skills and knowledge their trainees should have when they leave the classroom, they will be able to avoid drowning in trivia.

Gibberish

Sometimes the so-called instructional objective either contains or is composed entirely of phrases with little or no meaning. The following are examples of worthless expressions:

- Manifest in increasing comprehensive understanding
- Demonstrate a thorough comprehension
- Relate and foster with multiple approaches
- Have a deep awareness and thorough humanizing grasp

When these words precede a description of the desired performance, they just get in the way. And if no description follows them, the danger is more substantial. The danger is that people (other trainers and trainees) will be lulled into thinking something meaningful has been said.

They may even question their own intelligence trying to perceive a meaning that isn't there.

Although such verbiage may seem impressive, it is of little use in communicating instructional intentions. The best way to eliminate gibberish is to give the objective to a couple of trainees and ask them what they think it means. Their answers may be hard on the trainer's ego, but they will usually show the way towards a cleaner, simpler statement of intent.

Don't forget editors. A good editor can make miraculous moves toward simplicity and clarity by changing just a few words here and there.

Instructor Performance

Another practice that interferes with the usefulness of an instructional objective is that of describing what the trainers are expected to do, rather than what the trainees are expected to do. An instructional objective should describe trainee performance. It should avoid saying anything about trainer performance. To do otherwise would unnecessarily restrict individual trainers from using their best wisdom and skills to accomplish the instructional objective. When reviewing draft objectives, trainers should ask whether or not they are referring to trainee performance. If not, revise the objectives. For example, if we were to adjust the gear case objective to reflect instructor performance, it might read, "After reviewing the video presentation, have the trainees demonstrate the proper skills to rebuild a gear case."

Here, the objective restricts the trainer to using the video presentation, when another presentation method (such as a lab demonstration) might be more suitable.

False Criteria

A more insidious defect produces criteria that tells the trainee little or nothing that they don't already know. For example:

- To the satisfaction of the instructor
- Must be able to make 80% on the multiple choice exam at the conclusion of the course
- Must pass a final exam

The trainees know they must satisfy the trainer. What should be provided is a description of what they have to do to produce such satisfaction. If trainers do, in fact, make judgments about whether the trainees are or are not competent, there is no reason why those trainers cannot reveal the basis for their judgments. Of course that will take time and effort, but that's what professional instruction is all about. To test the criteria in an instructional objective, ask whether the criteria says something about the quality of performance the trainer desires, or says something about the quality of the individual trainee's performance. Individual trainee performance should be the focus. The standard should be real rather than vague or subjective based on the trainer's background.

The pitfalls to developing instructional objectives are summarized in Figure 5-3.

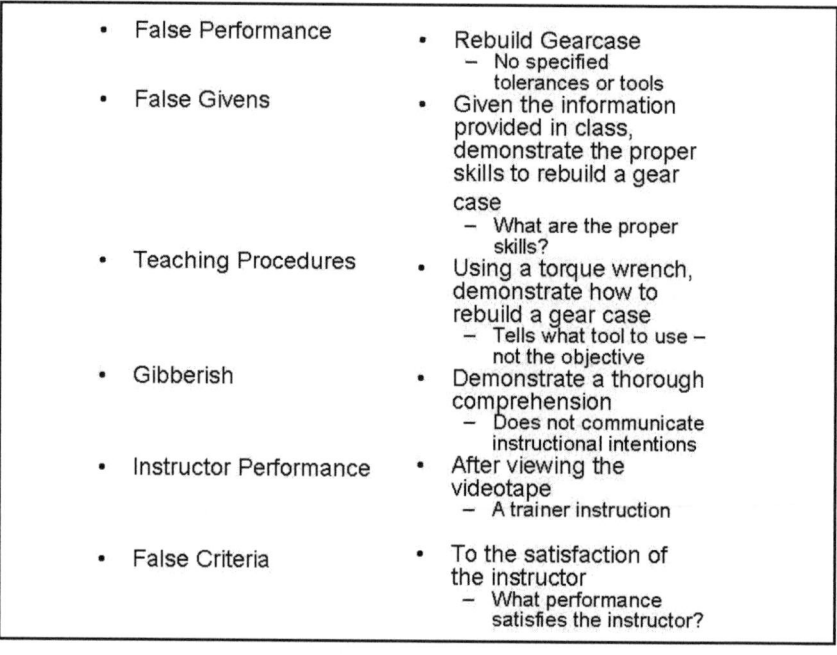

Figure 5-3 Pitfalls to Developing Instructional Objectives

Summary

An instructional objective describes an intended outcome of the instruction rather than the procedures for accomplishing those outcomes. An objective always states a performance, describing what the trainees will be doing when demonstrating mastery of the instructional objective. When the main intent of the objective is hidden, an objective clarifier through which the main intent can be clearly identified, should be added. Objective clarifiers are always the simplest and most direct as possible, and they are always something that every trainee already knows how to do well.

To prepare an instructional objective:

1. Write a statement that describes the main intent or performance expected of the student.

2. If the performance happens to be hidden, add an objective clarifier through which the main intent can be clearly identified.

3. Describe relevant or important conditions under which the performance is expected to occur. Add as much description as is needed to communicate the intent of the objective to others.

Checklist for Trainers

The checklist in Figure 5-4 serves as a useful summary for trainers developing instructional objectives. The checklist is divided into three

Performance
- Is your main intent stated?
- If the main intent is hidden, is an objective clarifier stated?
- Is that objective clarifier the simplest and most direct one you can think of?

Conditions
- Have you described what the trainee will be given, or be deprived of, during the performance of the objective?
- Have you described all of the conditions that will influence the shape of the performance?

Criteria
- Have you described how well the trainees must perform to be acceptable?
- Do these criteria describe some aspect of the performance, or the product of the performance, rather than the instructional process or meaningless percentages?
- When percentages are included in the criteria — does they reflect a realistic expectation?

Figure 5-4 A Checklist for Trainers

sections — performance, conditions, and criteria. This checklist will insure that the trainer gives proper consideration to these three main aspects of developing instructional objectives.

Once trainers understand how to develop the instructional objectives, the next step will be to develop or select the training material that will be required. This is the focus of Chapter 6.

Identifying and Developing Training Materials

Identifying and developing content are about closing the gaps between what the trainees already can do and what they can not yet do but need to be able to do. If there is no gap, then there's no need for training. There's no need to impose on the trainees (or the trainers) by trying to teach what the trainees already know.

If this idea were actually put into practice, it would be a pretty powerful one — the amount of training in the world could probably be reduced by at least half! So why isn't this idea put into practice? There are two main reasons:

The first is administrative convenience. Traditionally, we've developed the training and the lessons to fit into a predetermined amount of time, usually an hour or two. Because of this fixed time lesson, trainers have been forced to fit the training to the time, often stretching training or providing additional material, whether it was needed or not — as long as it fit the time period.

The second reason for excess training is courses have traditionally been content driven. That is, courses have been designed to teach as much content as the allotted time will allow. Because there is never any shortage of content, there is always enough content to fill the time period.

Training should be designed to meet a need, rather than being designed to fill time. Training should be designed to accomplish important outcomes. This change in approach should change the definition of a lesson.

Too often, a lesson consists of whatever amount of training fills the fixed time period. However, in some cases one trainee needs only 10 minutes to accomplish a particular objective, whereas another trainee may require two hours. But both trainees get the same fixed time period. During a traditional hour-long lesson, the first trainee wastes 50 minutes that could be spent on other objectives or returning to the job. Meanwhile the second trainee might master only part of the original instructional objective. Because time is fixed, performance levels achieved are variable — and in some cases time is wasted.

A better approach is for a lesson to consist of the training and practice required to accomplish that objective. The lesson should consume whatever time the trainee needs to reach mastery of that objective. To avoid confusion, this kind of lesson is called a module. It includes all the content and practice needed to promote mastery.

Selecting Content

Often there is more than enough content than there is either the time to teach it or an interest by the trainees in learning it. Therefore, the trainers must make some determinations about what content to include and what content to set aside. The first group — the content to include —is instructional material that falls into a "need-to-know" category. The remainder — the content to set aside —falls into various categories, including a) content that trainees need to know, but not in this lesson, b) content that would be nice to know, but is not essential, and c) content that is not necessary to know. How can proper content choices be made?

The trainers know what the important outcomes of the training should be and they know more or less what the trainees will be able to accomplish when they arrive. They also know what material is available in the location where the training will be held, and the restrictions under which they will have to work. Furthermore, the trainers know what information and experience they need to give the trainees to be successful. Therefore, developing the content for the training amounts to reviewing the requirements for achieving the instructional objective, the target population description, and the course hierarchy (see Chapter 5), and also answering the question, "what do the trainees need to know to demonstrate achieving the instructional objective now?

"What prevents the trainees from being ready to practice as soon as they enter the module?" The answer to that question will tell the trainer what needs to be done to fill the gap between what trainees can do now and what they need to be able to do in order to fulfill the entire objective of the module. The secret to deciding what to put in and what to leave out is to think about the difference between what the trainees already know and what they need to know. This difference can be expressed mathematically:

What needs to be known – What is already known = What needs to be taught

Why Aren't They Ready to Practice Now?

Sometimes trainees are prepared to practice as soon as they enter a module. Yet they still need training. Several reasons are possible. In some cases, there may be some aspect they don't know about how to do what they're expected to do. For example, they may know how to use a drill that's already set up, but not how to select the correct drill bit or how to insert and eject the bit. They may be some safety precautions for them to learn more about or common errors they should be able to avoid. Even if they know how to use the tools and materials, they may not know how to recognize what the desired performance looks like and feels like. Trainers will not want them to practice until they can monitor their own performance.

In other cases, the trainees aren't ready to practice now because they don't believe that what they're supposed to learn is valid. They don't believe the information really applies to their work. For example, those who don't believe the basic flow / pressure and speed arguments for hydraulics are not ready to hook up such circuits. Training — such as explaining the appropriate hydraulic principles — is needed before initiating practice exercises in order to convince them that flow and speed are directly related.

The Hard Part

The procedure for developing content is relatively simple. The hard part is getting used to the idea that the current lesson probably contains quite a bit more content and activities than needed to accomplish the objective.

The trainers all have their favorite topics, war stories, anecdotes, and demonstrations. They like the subject they are teaching and are all wrapped up in it. That being the case, discovering that some or all of what they do in the classroom can be better done without their favorite elements may be something of a blow to their egos. But if trainers are serious about making the training work — and work as efficiently as they can make it — then they need to think of those extras elements as obstacles rather than necessities.

Module Components

Before starting any content development, the trainers need to have a list of training module components. This list can be used to remind the trainers of the components that they should always include in a module as well as additional ones they might include.

Preparing the Training Module

These components are used to get trainees to the point where they are ready to practice the objective:

Basic components

1. The instructional objective
2. The skill check description
3. Description of relevance of training to the trainee
4. Modeling or demonstration of competent performance
5. Training content, including explanations and demonstrations
6. Alternative resources

Relevant Practice (Lab Exercises)

7. Practice
8. Feedback

Directions

9. Labels — module name and number
10. Directions to the trainees — "go get".
11. Directions to other trainers — "to teach this unit you'll need the fol lowing materials:"

Evaluation

12. Self checks
13. Skill check
14. Self-evaluation explanations

As can be seen, the basic layout of a training module is one that, at the front end, always informs trainees of the purpose and relevance of the lesson, and at the back end always provides practice, feedback, and evaluation. In between, it offers whatever training is needed to get the trainees from where they are at the beginning to where they are ready to practice.

How to Develop the Content

With this outline in mind, the trainers should be ready to consider the content development and identification procedure, which is pictured in Figure 6-1. The steps in this figure will help the trainers identify and developing only the specific material content that trainees require in order to complete the training module.

- **Step 1. Review the instructional objective.**
- **Step 2. Review the description of relevant practice for the objective.**
- **Step 3. Review the target trainees, material description and hierarchy. Note what the trainees can do if they are qualified to begin this module.**
- **Step 4. Answer the question, "Why aren't they ready to practice this module objective now" at the time the trainees begin working on the module.**

Figure 6-1 Content Development and Identification Procedure.

Imagine that a trainee has read the objective of the training module and understands the importance of learning what the module has to teach. Why would that trainee not then be ready to practice that objective right then and there? That's the question to answer. To make the question easier to answer, trainers can break the question into smaller ones.

1. Do you believe the trainees aren't ready to practice because they don't yet know some background information related to the subject that will be required to know how to perform the task properly? If so, what information do they need to know?

2. Do you believe trainees aren't ready to practice because there are one or more common errors they are likely to make in the present state of readiness? Which common errors are they and how will they avoid them?

3. Are they not yet ready to practice because they haven't been taught how to avoid certain dangerous situations? Which situations?

4. Are they not ready to practice because they don't yet know how to tell when their practice performance is satisfactory or not?

The answers to these questions will tell the trainers what content to include. If there is some standard content that isn't needed for answering these questions, leave that material out. When the trainers test the module, they will learn if content that was dropped from the module should have been left in. However, the reverse is not true. If the trainer puts something in that should be left out, testing may not expose the flaw. It's better to start lean, and add content and activities where necessary. The trainer should not worry about making the training too lean — they can always add more.

Here's an important tip on how trainers can complete this step of the development process. They should think of it as if they are developing a summary outline of the lesson content, rather than developing content. The content will be presented in an organized manner when the module is finished, but it can be an obstacle to begin filling in the content before the trainer has a summary outline. So the trainers should just list enough of the summary outline to answer the four previous questions.

If the trainers find that they only need to provide practice and feedback for some objectives, they should be pleased. The trainees will thank the trainers for refraining from boring them with information they already know. Furthermore, if the managers of the training program have the right focus, they will thank the trainers for getting the job done with a minimum of wasted resources. Trainers should never train just for training's sake.

Does this mean the trainees will get out early? Probably not, because they still have much to learn. However, the trainers can prepare a list of optional activities that trainees would find interesting and productive if they reach competence before the time is up. There are always advanced exercises that can be developed to challenge even the most skillful trainee. In fact, sometimes the other trainees will enjoy watching one of their own engage in an advanced exercise just to demonstrate their skills.

Trainers can use the technique described in this chapter for any course that they develop. However, they should be cautious about applying it to someone else's course. Although they will be able to identify all sorts of unnecessary training, nobody likes to be told that there is no need for some, or all, of what they are teaching.

Consider the example about bearings from Chapter 5, listed in Figure 6-2. As a training module is being developed for Duty 12-1, trainers would begin to examine the scope of the planned module. In order to keep this exercise to a manageable level for this chapter, we will focus on a training objective that states:

"Utilizing the training materials as a guide, the trainees should be able to safely remove and install a ball bearing on the lab unit on Stand 3."

---

12. Demonstrate knowledge and ability to properly install, inspect, or replace bearings. A list of tasks might be as follows:

12.1 Identify common bearing types and their applications.

12.1.1 Sleeve bearings
12.1.2. Metallic
12.1.3. Non-metallic
12.1.4. Ball bearings
12.1.5. Single row
12.1.6. Double row
12.1.7. Roller bearings
12.1.8. Cylindrical
12.1.9. Tapered

12.2 Demonstrate ability to properly mount and dismount bearings.

12.3 Identify the proper bearings seals for specific applications.

12.4 Demonstrate ability to properly lubricate various bearing types.

12.5 Identify various bearing wear patterns.

12.6 Demonstrate ability to find bearings in appropriate catalogs.

Figure 6-2 Duty- Task Analysis

Note: This training module is an advanced module. Trainees will have demonstrated that they can identify bearing types and find the bearings in the appropriate catalog.

If Step 1 from Figure 6-1 is to review the objective, the trainers are ready to move to Step 2. Now they develop the appropriate outline of the material and lab demonstrations that will prepare the trainee to safely remove and install a ball bearing on the lab demonstration equipment.

Once the trainers have prepared the outline and have a good idea of the lab demonstration that will be required, they are ready to move to Step 3. The trainers will need to consider who the target trainees are, what courses they have already completed, and what competencies they have been able to demonstrate. With this in mind, the trainers begin to select the "need to know" material to present to the trainees during the course. The trainers will also have to prepare a hierarchy (or order) in which the material should be presented.

With the material and demonstrations developed, the trainers will move to Step 4, which asks the question "Why aren't they ready to practice this module objective now?" By reflecting on this question, the trainers will be able to evaluate the course content, separating the "need to know" information from the "nice to know" information.

Content Development

Having completed previous training modules involving bearing identification, the trainees now enter the module that will teach them to remove and install bearings. They already know the difference between a ball and roller bearing, how they are constructed, and the theory of bearing operation. So why aren't the trainees ready to practice removing and installing bearings as they enter the module?

They're not quite ready to practice because they may not know how to position the bearing and shaft in a hydraulic press, how to press on the correct bearing race to prevent damage, where to safely stand during the operation of the press, or how much pressure they should use during the removal and installation. Are there any other reasons they aren't ready to practice? Probably none. But these points are what a module to teach this objective would contain. Consider this sample module:

1. The objective — This is what you should be able to do.
2. Performance check description — Here's how we'll check your competence.
3. Description of relevance — This is why this skill is important to the trainee.
4. Explanation of the procedure and examples of the final product of performance (demonstration of correct performance by the trainers)

—Here is how it is done, what it looks like when correctly done.

5. Practice with a series of objectives (recognizing correct perform ance) — This will help you keep from practicing the wrong thing.

6. Practice in performing, with feedback — Now it's the trainees' turn.

7. Skill check — Find out how well the trainees are doing.

Delivery System Selection

Having summarized the content of the modules, the trainers are ready to decide on how the training will be made available or delivered to the trainees.

This is probably the easiest part of the training development — that of deciding what combination of training delivery tools the trainers will use to present the training and practice to the trainees. Some would have the trainers believe this is a complicated affair, but it isn't. The first reason is trainers will not have so many choices available to them that they need to have a chart to decide which one to use. The second reason is by the time trainers have a list of the training delivery tools they need in order to provide practice and feedback, they will seldom need anything more.

Let's first consider the delivery system selection. Training experts often talk about delivery system selection rather than media selection. That may seem as though they're using big words when smaller ones will do, but there is a reason. Media are the message carriers: overhead projectors, chalkboards, computers, books, telephones, etc. They are the components on which the trainers write the information they want to deliver to their trainees.

However, trainers use more than these media to present their training. They often use people, either to present information, to participate in practice, or to assist in providing feedback. In addition, they often use job-related items such as machinery or equipment to assist with training and practice. Although these are critical requirements for proper presentation of the training, they are not media in the sense of the word. Thus, there is a preference for the term delivery system selection rather than media selection.

Advantages vs. Disadvantages

Trainers already know most of what they need in order to select suitable delivery systems for their courses. Trainers know the features of most of the available media (DVD's, videos, computer projectors, etc.), what they are used for, and how they should be used. Each delivery system has advantages and disadvantages.

For example, videotapes and DVDs can call up an image or moving sequence instantly. Does that make them the preferred delivery system? That depends on what the trainers are trying to accomplish. If they are trying to present a demonstration such as the proper way to hold a tool, or the sequence of steps for repairing a gear case, then videotapes or DVDs can be quite helpful. If the trainers are trying to teach students about smooth finishes, tight fits, or comparing weights among different nails, screws, or drill bits, the videotapes and DVDs are of considerably less value. The preferred method for this type of training would be tactile contact with objects, so the trainees can see and feel the differences.

This example shows that the features of any given delivery system are beneficial only when they help accomplish a purpose. If trainers match the benefits of each system with the specific goals of each lesson, they will avoid unnecessary razzle-dazzle of technology just for technology's sake, and help keep the training costs down.

In summary, there are at least six types of delivery systems, with multiple variations and combinations. Consider the list in Figure 6-3.

Delivery Method	Positives	Negatives
Lecture	Easy set up and delivery	Appeals only to the "Hearing" sense
White Board/ Flip Chart	Useful addition to Lecture – Trainer must write/ draw legibly	Must limit to small class size to be effective
Overhead Projector	Useful addition to Lecture – Trainer must write/ draw legibly	Must limit to small class size to be effective
Videos	Good for procedural examples	Requires projection/ playback set up and Trainer interaction
Computer Projector	Ultimate flexibility for training delivery	Requires technical set up
Lab Demonstration/ Hands-on	Good visual demonstration of tools/ techniques	Limits size to a small audience

Figure 6-3 Training Delivery Methods

Trainers may select a lecture in combination with a white board, a computer projector, and a video. They may be able to use a lab demonstration, with a flip chart (with some instructions for the lab), a short video (showing what is going to happen in the lab), or an overhead projector with images of cutaways that would be useful during the lab.

Each of the delivery systems has positive and negative features that trainers will want to consider before making their final choice. Also, equipment vendors have technical materials that may be useful in certain lessons. These materials may be in only one format, such as a video or computer image; the format might influence the selection of the delivery system.

Figure 6-4 highlights the typical utilization, audience size, and effectiveness (given size) of the delivery techniques. Although there are exceptions for some of these delivery techniques, this chart provides guidelines for their use.

Delivery Method	Typical Utilization	Size and Effectiveness
Lecture	Used to convey academic material	Effective for any size audience
White Board/ Flip Chart	Combined with lecture for illustrations and material emphasis	Audience size limits effectiveness
Overhead Projector	Combined with lecture for illustrations and material emphasis	Audience size limits effectiveness
Videos	Good for illustration of complex procedures	Can be effective with any size audience
Computer Projector	Can be used to incorporate all other deliver techniques – evolving to be the most popular delivery tool	Effective with any size audience
Lab Demonstration/ Hands-on	Good for hands-on demonstrations and allowing trainees to demonstrate proficiency	Small groups only

Figure 6-4 Training Delivery Utilization

Delivery System Selection Rule

The rule in selecting a delivery system is to choose the most readily available and economical items that will provide features called for by the training objectives. If trainers do not place the objectives first and foremost, they will be easy targets for sales representatives and marketing campaigns aimed at getting them to purchase expensive media hardware that they do not need.

Note: A common error is to decide on a delivery system for an entire course, rather than for a single objective or module. This decision is an error because the course could easily consist of different components, some of which could easily be learned by computer or other distance learning methods, others by one or more classroom sessions where direct trainer support is available. Other courses might be made up of different modules that would best be served by yet other combinations of delivery systems. The mistake is to think that a course has to be delivered by solely one medium or another, rather than by a delivery system consisting of a combination of methods and media.

Demonstrating Trainee Competency

Complete the following steps for each of your objectives:

1. List the tools, equipment, spare parts, and so forth that will be needed in order to provide relevant practice of the objective. To do this, the trainers would look at what they listed while describing relevant practice requirements. For example, to repair a gear case, one would need the proper tools, spare bearings, spare seals, replacement gears, etc.

2. Evaluate the items selected for practice. Focus on performing the tasks that will allow the trainees to make the most responses or get the most practice per unit of time. For example, it may be advisable for trainers to have the gear case partially disassembled before the training exercise. This preparation prevents the trainees from performing needless repetition of certain tasks and allows them more time to focus on more important elements of the exercise. They would now have time to concentrate on identifying defective or worn parts, proper alignment procedures for re-installing the bearings and shafts, and other key elements.

3. After the main points that need to be demonstrated are identified, time may be left in the lab or demonstration period for the trainees to engage in additional activities, especially if they are more advanced than some of the other trainees in the classes. If this is the case, the trainers will have to decide how to demonstrate these tasks without disrupting the remainder of the trainees.

For example, the trainers may provide lists of failure examples or descriptions of problem situations. They can have the advanced trainees either demonstrate how to solve a problem or write a short essay on how the problem should be solved. This activity will keep the advanced trainees challenged, while allowing the average trainees to complete the main objectives. The additional exercises are a challenge for trainers to manage, but are usually worth the additional effort.

4. If the training module requires academic content in addition to lab demonstrations or hands-on practice, then review the module content summary and determine how the trainer will present the content. The delivery method should have the best features for the content and objective involved. For example, when disassembling the gear case, are there certain mechanical fundamentals such as cleanliness that need to be stressed? If so, then content on contamination and its impact on the bearings, seals, and gears should be provided. Images depicting good and bad conditions could be included. A lecture on the impact that contamination has on the bearings could be supplemented with images of bearing failures; these images would re-enforce the points made in the lecture.

5. Once training has been developed and is being utilized for various trainee groups, the trainers need to consider the new trainees beginning the course. For example, what are their reading skills? If the trainers have decided to present information in print, but the new trainees have low reading skills, then the trainers should select a way to present the information that requires less reading, such as a progressive class discussion of the material. On the other hand, if the new trainees have strong self-study and reading skills, then various print and computer modules can be incorporated into the learning program. It may even be possible to establish an at-your-own-pace module, where the trainees work independently and use the trainers as guides or coaches when needed, and until it is time to demonstrate what they've learned.

Other factors include language, math proficiency, and skills with tools. Do any resources need to be available in languages other than English? Do the trainees all have similar abilities in math and calculations? Do the trainees need certain skills with tools in order to move forward, and do the participants have similar skills? All these factors need to be incorporated.

6. Are the tools and equipment the trainers have decided to use in the lab demonstration or exercise available? If not, trainers need to make other selections that will still allow the lesson objective to be achieved.

7. Are the tools and equipment that are selected for the lab exercise easy for the trainees to use, easy for them to present their demonstration, and easy for them to operate? If not, trainers should try to make more practical selections. For example, if there are multiple gear case and disassembly activities occurring at the same time, do the trainees each need a crane to assist with heavier components? Are chain hoists appropriate? Or another lifting mechanism? Safety and timely usage are important when considering the appropriate lab excercises.

8. Finally, could other lab demonstrations or exercises give the trainers the features they need, but at less expense and easier maintainability? If so, change the original selection. For example, suppose the trainers decided initially to present information by computer. A second consideration may convince them that a series of photos placed in a binder or on a display board, and accompanied by appropriate explanations, would be cheaper to produce and easier to maintain.

Now consider the actual lab equipment. If the trainees are to torque each bolt properly on a gear case assembly / disassembly lab exercise, how many bolts would the trainees need to replace? Because the bolts can only be properly torqued a few times, the fasteners would all need frequent replacement. If the trainees were required only to demonstrate properly torquing one fastener per exercise, the number of replacement fasteners would be reduced.

These steps should complete the selection process. There may be times when decisions can be a little trickier to make, as when trainees are scattered around the world. But most of the time, the trainers will simply need to make selections that provide or enable relevant practice and feedback.

The Lesson Plan

A lesson plan is a prescription for training, a blueprint describing the activities that trainers and trainees may engage to reach the objectives of the course. The plan's main purpose is to prescribe the key events that should occur during the module. If trainers find it necessary to deliver most or all of the training through lectures, the lesson plan is a guide to their actions. When the module is put in the hands of the trainees, the plan performs a similar function: it tells them what to do, where to locate the training resources, how to practice, and how to demonstrate competence when they are ready.

The precise format of the lesson plan is less important than making sure it performs its important functions. By now, trainers have already listed or summarized the content of each lesson. Therefore, the task of preparing the module will be relatively simple. Whatever the format used, however, it should emphasize the trainees rather than the trainers. That way the trainers will avoid the trap of developing courses based on their own preferences, instead focusing on what the trainees need to accomplish the course objectives.

Try Out the Lesson

One mark of a professional is the insistence on tryouts before going public. Many Broadway plays are developed out of town, nightclub acts may be polished in the smaller lounges, and consumer products are tested until they achieve customer satisfaction. Similarly, training and job aids such as questionnaires and surveys are often tried out before being considered ready for regular consumption. The time that trainers spend on tryouts is time they will never regret.

Trainers will often receive substantive and cosmetic comments, leading in turn to substantive and cosmetic revisions. What is the difference? Substantive comments suggest important changes to the content or the sequence of the content. For example, suggestions that trainers should move or reorder the modules, delete unneeded material, or correct technical errors are substantive changes. Keep testing the material until the number of substantive comments has dropped to zero — or where the remaining comments representing areas where the trainers deliberately agree to disagree with the recommendations.

Cosmetic comments refer to style. When individuals evaluating a module comment on the trainers choice of words, their manner of writing, or their political correctness, the trainers are hearing comments about the cosmetics of the course. These are comments the trainers also want to pay attention to because they affect how the trainees will receive and respond to the information. For example, sometimes the content is accurate (substantive), but the presentation is unclear because of weak sentence structure (cosmetic). Clarity of style can be as important as clarity of content in helping trainees understand the information being taught.

Sequencing

Not everything can be learned at once. Therefore, training must be offered in some sort of sequence. One lesson must come before another. But there doesn't always have to be a prescribed order; that is, the lessons don't always have to be studied in the same sequence by each and every trainee. For example, if the course involves a lot of math steps, then the order is often more important. One skill becomes prerequisite to another. On the other hand, suppose the course involves learning to work with a variety of different hammers, and there are different stations for each hammer. Here, the sequence may be less important.

If the trainers follow the "how to do it" steps described in this chapter, they will find the job of drafting their training greatly simplified. To make their training work, trainers should make sure they teach information and skills that the trainees do not yet know. Trainers should provide the training needed to get trainees ready to practice the objective, provide feedback and practice, and then take steps to find out whether the trainee can perform as desired. If this process is followed, the instructional objectives will be achieved consistently.

Training the Trainers

In 1996, an American Society for Training and Development study showed that U.S. corporations spent more than $55 billion in formal employee education and retraining. Ten years later, they spent $129 billion on employee training. Of this $129 billion, internal training accounted for $80 billion, and external training for the remaining $49 billion. Given that the majority of total training expenditures are spent on internal training, the internal trainers take on increased importance.

As noted in Chapter 1, many pressures are put on companies trying to cope with the shortage of skilled workers. These pressures, combined with rapid changes in technology and continued shortcomings in the U.S. educational system, intensify the challenges on internal training organizations. When a company's training organization tries to help the company respond, one of the areas where improvement can be generated is with the company's own group of internal trainers.

Highly-skilled trainers can have an enormously positive impact on the workforce. Thus, the company's training organization must focus on selecting the best trainers for the company's needs. Where do highly skilled trainers come from? How can a company insure that trainers continue to be highly skilled? How can their positions be set up so that trainers find continuing challenges in the workplace that will enable them to always be motivated?

What Is a Trainer?

The starting point to address these questions is to answer what may be a self-evident question at this point: What is a trainer? For our purposes, a trainer is anyone who helps people to increase their knowledge or their skills. There are many different types of trainers —work management trainers, people management trainers, personnel management trainers, and so forth. This book will focus on trainers for technical skills.

Is a trainer a facilitator or a teacher? Typically we think of trainers as people whose primary focus is on relaying content — technical or otherwise. Beyond that, we tend to see a facilitator's main focus being on learning through group processes. A teacher, in turn, generally brings to mind someone whose main teaching method is lecturing and who is responsible for discipline. Human resource professions who specialize in development and performance fields often have strong feelings about the differences between facilitators and teachers. They view teachers as the person taking the lead in stand up lecturing and keeping the class orderly. Facilitators take an agenda and a group of adults, then draws the group to a consensus agreement or understanding. Facilitators rarely "teach" or interject their own knowledge or opinion on a subject.

To many, good trainers are both good facilitators and good teachers. Because trainers are found in the work environment, they work primarily with adults. Therefore, good trainers effectively and efficiently use a variety of appropriate methods to help a broad range of adults acquire new skills and knowledge; thus, they are teachers. However, good trainers are also good facilitators when they can help the trainees capitalize on the knowledge they already posses and be guided toward a solution to a problem.

Learning How to Train

Trainers need to master four essential areas of basic knowledge and skills. These are:

1. Learning theory
2. Training styles and methods
3. Presentation and delivery
4. Evaluation

Chapters 5 and 6 discussed methods for preparing instructional objectives and developing training materials. In this chapter, we are assuming the trainers will be delivering material that is already developed according to the methods of these earlier chapters.

Technical trainers are typically subject matter experts who are called upon to train. But simply having a skill or specialized knowledge does not automatically mean that a person can train others in that skill or convey their knowledge. Trainers must first master a set methodology that their organizations need in order to train others as effectively and efficiently as possible. This translates into a focus on the instructional objectives, the delivery methods, the desired performance outcomes, and the job duty perceptions that trainees need to acquire.

Successful trainers have two key personality qualities. The first is presence, which is the ability to hold people's attention and be taken seriously. The second is self-confidence, which allows trainers to handle skepticism or rejection from those inevitable hostile participants who for one reason or another do not want to be in the training.

In addition to technical skills, trainers must also understand training theory and instructional program development. Therefore, the trainers themselves must be trained in the four areas listed above in order to be effective.

Learning Theory

Even though a training program may already be developed in terms of objectives and content, within each program, there is a lot of flexibility that the instructor can bring to the course. This means that trainers will need to be familiar with the following:

1. Effective ways to present the material
2. Dealing with problem participants
3. Keeping momentum going
4. Defining what the participants have learned

Trainers must understand the difference between training adults and training children and adolescents. Adult trainees have a vast amount of experience from which they have learned; they are responsible for making their own decisions and living with the consequences. Thus, adult trainees need to see the relevance of any training to their own life experience. This is particularly true because most will be technicians who are already working as apprentices in a maintenance department.

Adult trainees learn best when they have some control over their own learning experience. They regard growth and self understanding equally as important as growth and learning, which make teaching "why this material is valuable" very important. They also prefer to take an active part in the learning process. In that regard, Adult trainees excel in task- or experience-oriented learning situations — supporting the importance of hands-on lab exercises — as well as cooperative climates that encourage risk-taking and experimentation.

Nevertheless, the most common learning style involves trainers lecturing in front of the trainees, who absorb the information imparted from the trainers. Even within this traditional setting, the trainers must keep in mind that adult learning is most effective when the trainees can satisfy personal goals or needs. Many adult trainees respond to extrinsic factors such as promotions, job changes, or better working conditions. But intrinsic motivators are just as important; they include self-esteem, recognition by peers, better quality of life, greater self-confidence, and the need for achievement and satisfaction.

Trainers should understand adult trainee learning styles and why some techniques may work better in some situations than others. Adult trainees gain information in two ways:

• Actively — through direct involvement
• Passively — through absorption of information

Furthermore, adult trainees process the information in one of two ways:

• Deductively — from the general to the specific
• Inductively — from the specific to the general

The ultimate objective of training is to increase performance through a change in behavior. Adult trainees generally go through four levels of learning to reach the level of behavioral change. These four levels are:

1. **Awareness.** The learner experiences the learning situation or event.
2. **Understanding or knowledge.** The trainee places the learning event in context, connecting causes, components, and consequences associated with the event.
3. **Skill.** The learner applies the understanding or knowledge learned.
4. **Attitude.** The learner sees value in the application of new knowledge and skills.

At the skill level, the trainee can perform the new behavior whereas at the attitude level, they want to perform the new behavior.

Training Styles and Methods

The classical style of teaching — the instructive style — is generally regarded as trainer led and subject centered. Many training experts now believe, however, that the facilitative or participative training style is more appropriate for adult trainees. In this style, trainers guide the trainees to discover what is to be learned. It is defined as trainer facilitated and trainee centered, better suited in most cases for the adult learning style.

Most adult learning situations are better suited to the facilitative or participative style, yet the instructive style still has its place under certain circumstances. When trainers are aware of the differences in learning and teaching styles, they can deliberately apply what is most appropriate for the learners and the situation.

Trainers who are knowledgeable about adult learning theory can apply that theory to their style and method of training. All predesigned training programs should designate appropriate training methods throughout the course. It is important, however, to know the variety of training or facilitating methods available to use according to the trainees' needs and the information that has to be learned. The following discussion looks at some widely-used training methods.

Lectures

Lectures are undoubtedly the most popular training method. Trainers stand before the trainees and deliver the information to be learned, whether solely by speaking or by combining speaking with a variety of audio and visual support such as slides, recordings, overhead projector, and brief videos. Properly-designed lectures can impart a considerable amount of information to various groups of trainees. However, they require superior presentation skills from the trainers and they limit audience participation.

Group Discussion

Group discussion is an informal training method with a leader or moderator who guides the trainees to share information and experience. Discussions need to be well-organized and limited to small groups. Trainers and trainees all benefit from the opportunity to contribute to the learning activity, but this teaching method can be time-consuming. Sometimes trainers will encounter trainees who are either unresponsive or disruptive. These trainees will be those who talk too much, or not enough, or who monopolize the discussions.

Readings

In this method, assigned readings contribute to the trainees' store of knowledge. Readings can augment information acquired through the lecture or discussion. For maximum effectiveness, the trainees should be encouraged to analyze what they have read.

Role Play

In role play, trainees are assigned predetermined roles to act out in a given situation. The purpose is to have the trainees learn how to solve a problem or achieve a level of understanding of people in roles other than their own. Role play is not appropriate for large groups, trainees who may feel threatened, or those who are too self-conscious to take part. Role play can be used when differentiating between the jobs of a maintenance planner, a maintenance supervisor, and a production manager. The role play could be structured so that there are real world conflicts that have to managed and overcome to accomplish a maintenance task.

Simulation

In some cases, simulations are "games" that have been developed to show how the roles of certain employees impact the overall business of the company. In other cases, simulations are detailed computer programs that allow the trainees to perform in a virtual environment, controlling a computer figure. They can make mistakes and see the result of the mistake in a virtual environment. This allows for corrective instruction to be given by the trainers or, in some cases, the computer program.

Games

Games incorporate competitive activities governed by rules that define trainee actions and determine outcomes. Games demand a high level of trainee involvement, while facilitating meaningful and fun learning. They can also be used as icebreakers or warm-ups for new groups.

Panels

In this method, a moderated group of 3–5 experts read prepared statements related to a chosen topic. They then discuss that topic with one another and respond to audience questions. Panels provide different opinions and thereby provoke better discussions. The frequent change of speakers keeps the attention from lagging.

Demonstrations

In a demonstration, trainers show trainees how to successfully complete a given task by performing the task themselves, adding supporting explanations and commentary along the way. For example, the trainers may disassemble a gear case and make appropriate comments and observations as they do so. Demonstrations stimulate interest and engage the trainee's attention. They should be carefully planned and limited to small groups.

Case Studies

In this method, a statement, a problem, or a case study is often followed by a group problem solving session. This is like a problem-solving or a brainstorming session, where the trainers present the case study and facilitate the trainees' responses until the correct answer is derived. To find a solution for complex issues, this method allows trainees to apply that new knowledge and skills they have acquired through other teaching methods.

Presentation and Delivery

No matter how many participatory exercises are built into a training program, the trainers will eventually need to fill the role of lecturer or presenter. Sometimes trainers fail in their objective of changing behavior or attitudes simply because their presentation styles fail to grab their audience. However, some trainers simply have "it" — the ability to capture and hold the audience's attention. Fortunately, many of the features that make up "it" can be learned. This discussion considers some of the tips on achieving the kind of presence that grabs and holds an audience.

Use Effective Openings

The opening sets the tone for the presentation and can make or break the training session. A good opening will succeed in the following ways:

1. Capture the audience's attention.
2. Reveal the trainers training style.
3. Raise the comfort level of the audience.
4. Introduce the topic of the presentation.

A number of techniques can be used to warm up the audience. For example, the trainers may share a personal anecdote related to the subject. They may ask questions relevant to the subject. Effective trainers may use creative introductions. Rather than simply going around the room having trainees introducing themselves, they may have them introduce one another or respond to specific topics such as "the most important book I have read" or "what I hope to learn from this course" — some way that requires each member of the group to participate.

Set Expectations

Let the trainees know what to expect from the beginning. Basic information can be worked into the opening or presented immediately afterward. Trainees will need to know the following:

- Who the other trainees are — either directly or through techniques described earlier
- Objectives of the session — what they will learn and why it is important
- Instructional techniques — how they will learn, e.g., discussions, videos, lecture
- Evaluation expectations — how their learning will be evaluated and how they will critique your presentation area
- Agenda — times for sessions, plus any assignment schedules

Structure Your Presentation

A trainer's job often calls for presenting a structured lecture. The following four elements provide structure to an effective lecture or a presentation.

1. **Introduction.** This element should get the trainees' attention, introduce the key points of the presentation, establish a rapport with the trainees, state a benefit to be gained by the trainees, and create anticipation for the rest of the presentation.

2. **Bridge.** This element is the transition to the key points, often by way of example or anecdote.

3. **Main Body.** This element provides the heart of the discussion supported by facts, figures, and examples. It also includes the rationale for the presentation.

4. **Close.** One of the most important elements of the entire presentation, the close should paraphrase key points, restate the most important point, and contain a re-statement that tells the trainees the benefits for them of the presentation.

Develop a Presence

The quality of the trainers' presentations can be developed. A good trainer's presentation will include the following qualities:

1. A pleasant, appropriate appearance
2. Effective use of body language and natural, open gestures
3. A well-paced delivery style, with effective use of pauses for emphasis and reinforcement
4. A well-pitched voice, loud enough to be heard well without being grating or annoying.
5. Genuine enthusiasm and sincerity
6. Effective eye contact with the audience
7. A natural, relaxed style that puts the audience at ease

Evaluation

An evaluation of the trainees should be conducted on four levels. These levels are:

1. **Awareness or Reaction.** Are the trainees happy with what they're getting? Is the material relevant — Is the training design appropriate? How effective is the trainer's leadership?
2. **Knowledge and skill.** Do the materials and methods actually teach the attitudes, concepts, and skills as they should?
3. **Behavior.** Will the trainees change their behavior based on what they have learned? Will they use the newly-acquired skills, attitudes, or knowledge back on the job?
4. **Results.** Are the new behaviors having (or will they have) a positive effect on the organization?

Generally, evaluation becomes more challenging as you proceed from the first level to the fourth level. To compensate, evaluation instruments usually are built into the training design, especially for the second through the fourth level. Trainers generally administer the evaluation, especially to get reactions and measure how much has been learned.

Techniques for Level One Evaluation (Awareness or Reaction)

At this point, we will discuss only Level One evaluation. It is the easiest as it is usually part of an ongoing process. At Level One, trainers will probably use some kind of predesigned questionnaire that asked for feedback related to various aspects of the program, such as length, content, lectures, audiovisuals, training methods, handouts, organization, facilities, and trainer facilitation.

Here are steps to follow when performing a Level One evaluation:

1. Ask the trainees to say three things they have liked about the pro gram. Then ask for three things they would like to see changed for the next training group.

2. Paraphrase what each trainee says; list their comments on the card or board. If you need more clarification, simply ask for more detail: do not defend or argue. Be sure to thank the group for their input.

3. As an alternative, you can hand out sheets with the questions and spaces for three likes and three things the trainees would like changed.

Then,

- A. Ask the trainees in small groups to discuss their answers.
- B. Have one member of each group report to the rest of the participants.
- C. Put the responses on a flip chart or board.

Steps A and B will take about five minutes. Step C takes an additional five minutes or so.

This technique is used for midcourse evaluation and for building trainer credibility with the group. It is often best used just before lunch break or at the end of each day. Do not use it unless you genuinely want to hear what people have to say about the program or the trainer — even if their feedback is negative.

Selecting Trainers

An important task for the training department is selecting people to be trainers. This task includes the following steps:

- Identifying the skills the trainers must have, such as communication skills and presentation skills
- Describing additional characteristics that it would be good to have, such as managerial skills and writing skills
- Considering the trainers' career paths, including lifetime trainer, managerial aspirations, and consulting

Selecting good trainers is essential to successful training. A mistake can be expensive, time-consuming, and disruptive. Therefore, you need to establish a good interview and selection program. Start by answering questions such as these:

- What characteristics should trainers have?
- What specific skills should they have?
- What technical credibility should they have?
- What experience should trainers have before being considered?

This list can grow very quickly. Separating the question into the seven categories pictured in Figure 7-1 can make the list manageable.

Categories of Trainer Evaluation

- Education
- Experience
- Job knowledge
- Interpersonal or people skills
- Knowledge of the training process
- Trainer skills
- Career goals

Figure 7-1

You may need more or less categories than these. However, you can take each category and further define it. This process is highlighted in Figures 7-2 through 7-8. Each of these figures expands on one of the categories from Figure 7-1. These checklists may be used as guides to developing your own checklists.

Education

- Is a formal education required?
  - How much?
- If a degree is necessary, which one?
- What knowledge is necessary?
  - Is graduation from the apprentice program necessary?
  - Is a journeyman status required?

Figure 7-2

Experience

- Are the trainers qualified to perform the subjects that they will be instructing?
- Do the trainers have any experience training people?
- Have the trainers ever made a presentation?
- Do the trainers have any experience in designing training programs?

Figure 7-3

Job Knowledge

- How much job knowledge is required?
  - Craft Specific?
- Do they have the ability to present theoretical and practical material?
- Can they grow into the role of a trainer?
  - If their skills are not complete at the start

Figure 7-4

Interpersonal or People Skills

- Can the trainers function in a classroom and in a lab?
- Can the trainers:
  - Empathize with trainees?
  - Get their message through to the trainees?
  - Handle conflicts?
- Are the trainers:
  - Egotistical?
    - Can they suppress it to get through to the trainees?
  - Independent?
  - Confident?
- Do the trainers relate well to the trainees?

Figure 7-5

Knowledge of the training process

- Can the trainers prepare instructional objectives?
  - With some training?
- Do they understand the role of a Trainer?
  - With some training?
- Do the trainers understand learning styles?
  - With some training?
- Do the trainers understand group dynamics?
  - With some training?

Figure 7-6

Training Skills

- Do the trainers have good verbal communication skills?
- Are the trainers good facilitators?
  - With some training?
- Can the trainers coach?
- Can the trainers read a group of trainees and adapt their training style to fit the group?
- Can the trainers critique and also take critiques?
- Can the trainers keep information confidential?

Figure 7-7

Career Goals

- **Do the trainers or should the trainers want to be trainers?**
  - Are they better suited as future managers?
- **Does being a trainer really fit into their personal career goals?**
- **How long will they want to be a trainer?**
- **Would the trainers be a candidate to be training counselors or managers?**

Figure 7-8

Once the checklists are complete, place the information into three additional categories, from which you can develop a matrix. The three categories are

- Must have skills
- Good to have skills
- Future desired skills

This information can be developed in the matrix style seen in Figure 7-9 on the following page.

A Trainer Qualification Example from the Case Study

In the training program that was described at the start of Chapter 3, the trainers were journeyman technicians who were selected from over 1500 maintenance technicians at the plant. When the training program first began, each area's maintenance department nominated one employee that would make a good trainer. Each of these 40 employees was evaluat-

Trainer Decision Matrix

Category	Must Have	Good To Have	Future Desired Skills
Education			
Experience			
Job Knowledge			
People Skills			
Training Knowledge			
Training Skills			
Career Goals			

Figure 7-9

ed by the human resources department. They were also interviewed by the training department manager, the training coordinator, and the maintenance superintendent.

From these 40 employees, the original 10 trainers were selected. These trainers would be responsible for developing the apprentice training program and for instructing the classes for the first group of apprentices. The union and plant management had an agreement where the trainers would retain their status as hourly employees and their seniority in their original departments.

The ten trainers needed approximately two years to develop the apprentice training program to the point that it could begin. The trainers were provided time for ongoing program development as the training program progressed. When the first program was finished, the trainers were given a choice to stay with the training program, or to return to their original assigned maintenance department, where they would be next in line for promotion to a supervisory position.

If they stayed in the training department, they would continue to gain tenure as an instructor. They then committed for two more years to see the next group of apprentices through the program. At the conclusion of the second apprentice class, they were then given a final opportunity to return to their original maintenance department.

If they did not return to their department at that time, then in most cases they would not be offered another opportunity to move into maintenance supervision. If in the future, they decided to leave the training department, they could return to a journeyman technician position in their original department, but most likely would never be considered for a supervisory position.

If trainers chose to stay with the training department, they still had good assignments. They were paid their journeyman's rate, which was Zone 12, the maximum pay rate for a journeyman technician. When they taught, they were paid Zone 13, which was the pay rate for a lead person. Most trainers' load was 20 hours of direct classroom training per week, and 20 hours of preparation or development time per week.

One night per week, the trainers would be required to stay 2–4 hours for what was termed a "help session." These sessions were designed to provide one-on-one coaching and mentoring for any students having difficulty with the material. Classes typically ran from 5:00 PM to 9:00 PM, with one electrical and one mechanical instructor available each evening. The trainers would make a top hourly rate and have four hours overtime each week. The trainers also had the option, if they chose, to make themselves available to work overtime in their original maintenance department. The overtime would not be allowed to interfere with any of their training schedule. However, they could work part of the second shift from 5:00 PM to 11:00 PM or they could work on weekends.

In addition, several vocational schools in the area were aware of the apprentice training program. They would frequently contact the trainers in the apprentice training program and work with them to set up various vocational programs that could be taught in the evenings. This arrangement was permitted by the company and again allowed the instructors an additional source of income. In addition, several of the instructors had finished their college degrees and were used to teach evening classes at the local university. All in all, therefore, being a trainer could be a very positive and financially rewarding experience. Thus, many trainers chose to forgo supervisory positions and become permanent trainers.

After the original 10 trainers had been chosen, and the first group of apprentices graduated, a pool of potential instructors was developed. To be eligible to become an instructor, the candidates had to meet certain criteria. The first was potential trainers had to finish in the top 10% of their graduating class academically. The second was they would need one or more of the current trainers to sponsor or recommend them to the training manager as candidates for trainers. The current trainers would typically observe apprentices as they progressed through the program, watching for those who were good communicators, those who would work well with others during training, and those who had a self-confident attitude.

Once approved, the potential trainers would be brought into the program as lab assistants; they would work with one of the existing trainers. At first, they would set up the lab exercises and help the leading trainer coach and monitor the class lab exercises. As they progressed, they would take the lead in coaching and monitoring the lab exercises, with the lead trainer working as their assistant. As they continue to progress, they would conduct 15–30-minute segments of the classroom material, with the lead trainer observing and coaching. Finally, when there were openings for instructors, the lab trainers would move into the lead training positions. This constant progression always ensured that a quality trainer was available for the apprentice program.

Reflections for Technical Trainers

The object of any training session is to make a difference that matters. Unfortunately, a chronic problem that trainers face is relevance. It's often hard for trainers to tell how much difference, if any, they are making. Is the training relevant enough that the trainees will actually use it on the job?

Relevance is rooted in a much deeper problem: The training paradigm in most companies is still too attached to the Industrial Age, which views people as interchangeable parts in the great enterprise machine. This kind of thinking treats participants like computers, and trainers like data entry personnel.

Of course, some types of training — new technology, new office skills — will always be instructional. But it's imperative that trainers don't continue to apply an outdated approach to teaching higher-order skills such as decision making and problem solving. The challenge for trainers is to banish the Industrial Age mindset and recognize that people are not things. They're enormously gifted individuals, each uniquely capable of contributing more than they are asked to contribute.

One key question every organization should ask itself is, "Do you believe that the majority of people in your organization possess far more talent, intelligence, capability, and creativity than their present jobs require or even allow?" Almost every organization will answer "Yes!"

In this regard, the challenge that trainers face is to lead people to new insights and inspire them. Trainer should carefully consider what key insights they want participants to gain and what paradigms must change. Trainers need to discover what excites and inspires them (the trainers) about the training, and how that inspiration touches on a higher mission or deeper values — for example, changed behaviors that improve business results, such as eliminating waste. With this discovery, real learning starts.

Trainers help others translate their knowledge into action. They should help the trainees align their own talents and passions so that they can make the changes necessary to the task at hand. Trainers should challenge them to define individually how their new understanding translates into new behavior. It's one thing to have new insights and new goals; it's quite another to put them into action. As a result of good training, trainees should complete their sessions with a clear idea what to do differently.

As they develop and delivery training, good trainers will always keep in mind their business's greatest areas of pain.

- What keeps your executive management team awake at night?
- What business issues are currently giving them their biggest headaches?
- What do they see as their biggest source of concern over the next few years?

Facing constraints such as downsizing and tight deadlines, most training professionals will need to do more with less. They will have to work with reduced budgets, faster turn-arounds, and a shortage of subject matter expertise. Training organizations must find ways to use their limited resources to deliver highly repeatable, scalable training sessions. Utilizing many of the hints and tips in this chapter should help technical training instructors add value to their organizations.

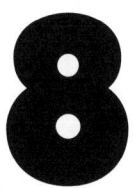

EFFECTIVE CLASSROOM CONDITIONS

Endless options and equipment combinations are available when setting up effective classroom and lab training conditions. It would be beyond the scope of this text to outline all of these combinations. We will focus instead on the lab-classroom set up that was used in the case study from Chapter 3. Starting with this information as a foundation, you should be able to determine the ideal lab and classroom setups for your training program.

Consider first the overall training facility layout for the program. Figure 8-1 shows the overall layout for the classrooms and labs. The series of rooms were built inside an existing plant building. The classrooms listed 1 through 8 were separated from the classrooms 9 through 12 by an exterior building wall. Classrooms 9 through 12 were actually in a courtyard surrounded by existing buildings. Examining the figure more closely, Classrooms 1 and 2 were the trades and craft combination lab and classrooms. Classroom 3 was the main mechanic's classroom. Classroom 4 was the mechanical lab. Classrooms 5 and 6 were general classrooms where various mechanical and fluid power topics were taught in a lecture format. Classroom 8 was the hydraulics and pneumatics lab. The room marked 8 was actually not a classroom, but rather the offices for the four mechanical instructors. Classroom 9 was the AC electrical classroom and lab combined. Rooms 10 and 11 were offices for the electrical instructors. Classroom 12 was the DC electrical classroom and lab.

This layout was convenient, and allowed for movement between classes without a lot of lost time and with very little disruption to other classes that may have been ongoing. The distance from the doors to the mechanical area to the electrical classrooms was only about 75 feet, which was traversed rather easily even during inclement weather.

Figure 8-2 details the layout of the trades and craft classroom and lab. The classroom and lab consisted of an open area for most of the space, with the exception being the wall between the blocks marked "A" and the

Classroom Layout

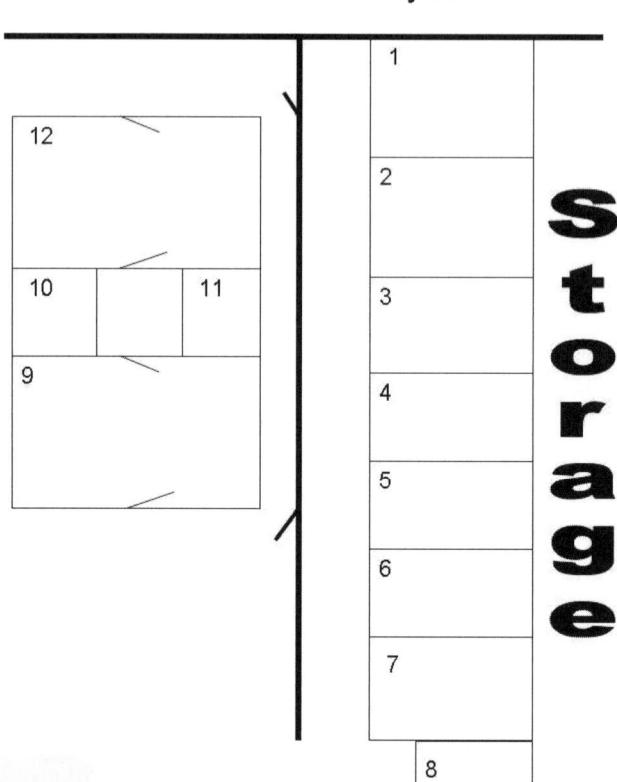

Figure 8-1

rest of the classroom. The reason for the separate wall was the blocks marked "A" were burning stations with an oxygen acetylene cutting torch hook up. The wall prevented any interference between the trainees who were practicing with the burning torches and the trainees performing other activities in the lab.

The equipment was dispersed throughout the lab so that multiple activities could be performed at the same time, allowing the trainees to work safely. For example, the blocks marked "B" were pedestal grinders. The blocks in the diagram marked "C" were drill presses. The block marked "D" was a cut off band saw. The block marked "E" was a pipe machine for threading pipe and studs. The spacing of this equipment was such that the trainees could be working on each piece of equipment with-

Trade and Craft Layout

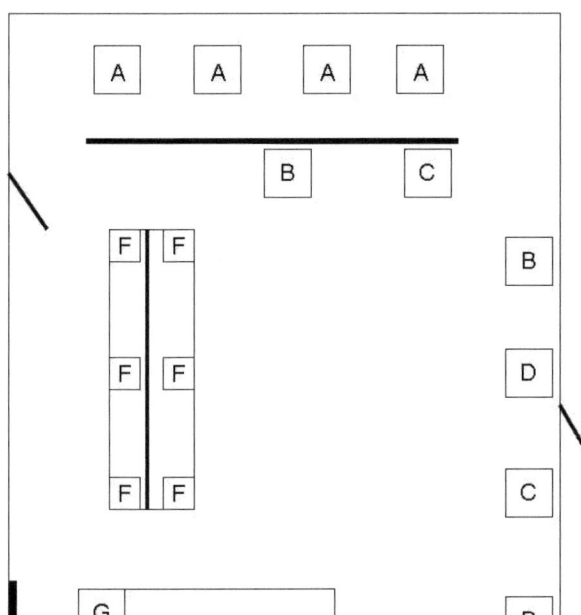

Figure 8-2

out interfering with another trainee.

The table marked "G" was the classroom table. Up to 12 trainees would sit around the table for any general training, such as math or blueprint reading. An overhead projector and computer projector were available and the images would be displayed on the screen at the head of the table. The large table with the six "F" blocks was a large workbench. The line through the middle of the workbench represented a vertical panel that extended from one end of the workbench to the other. This panel separated the work areas.

The panel was also used as a shadow display board for the various types of tools that the trainees would use throughout the class. Each tool was shadowed on the board and properly labeled, which allowed for easy

Mechanical Classrooms

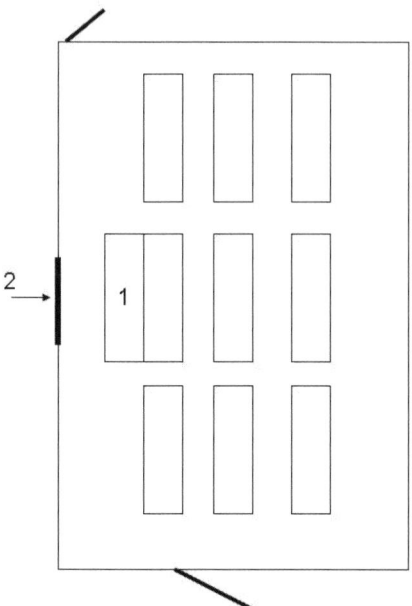

Figure 8-3

identification by the trainees. At the position marked "F" was a vice the trainees could use during shop work. Considering the types of exercises that the trainees were required to complete, as described at the start of Chapter 3, this combination of classroom and lab was very effective.

Figure 8-3 shows the typical layout for the mechanical classrooms, although from time to time the trainers would change the classroom from a portrait layout to a landscape layout. Other than rearranging the tables and the projection screens, the classrooms were virtually identical. In Figure 8-1, the classrooms that were arranged as shown in Figure 8-3 were rooms three, five, and six. The tables in these classrooms were heavily constructed and could be used to support models and actual equipment components for instructor demonstrations. Each of the classrooms was equipped with an overhead projector, a computer projector, and a screen. The computer projectors were all capable of video projection, whether from a VCR or a DVD player. This allowed the instructors flexibility in

Hydraulics and Pneumatics Lab

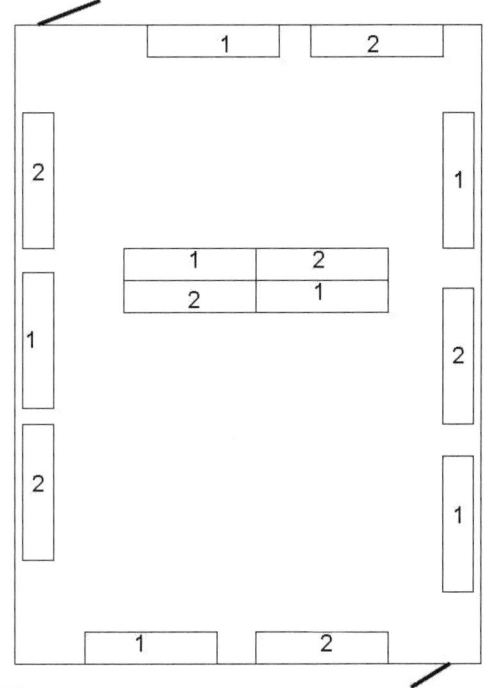

Figure 8-4

how they would present the material. Each of the classrooms also had a front and rear exit for safety.

For larger demonstrations, and even for some training exercises involving component disassembly such as pumps and gear cases, Classroom 4 was used (see Figure 8-1). In addition to heavy tables, this classroom had several chain hoists available to make component disassembly easier.

Figure 8-4 shows the details of the hydraulics and pneumatics lab. In this lab, all of the blocks marked "1" were hydraulic workstations. All the blocks marked "2" were pneumatic workstations. The configuration for both sets of workstations was very similar. There was a vertical 1/2" thick 2' x 4' steel panel attached to each station. Various hydraulic and pneumatic components were front mounted on these boards. All of the components had front mounted quick disconnects attached to them. This would allow the trainees to "hose up" virtually any hydraulic or pneumat-

AC Lab

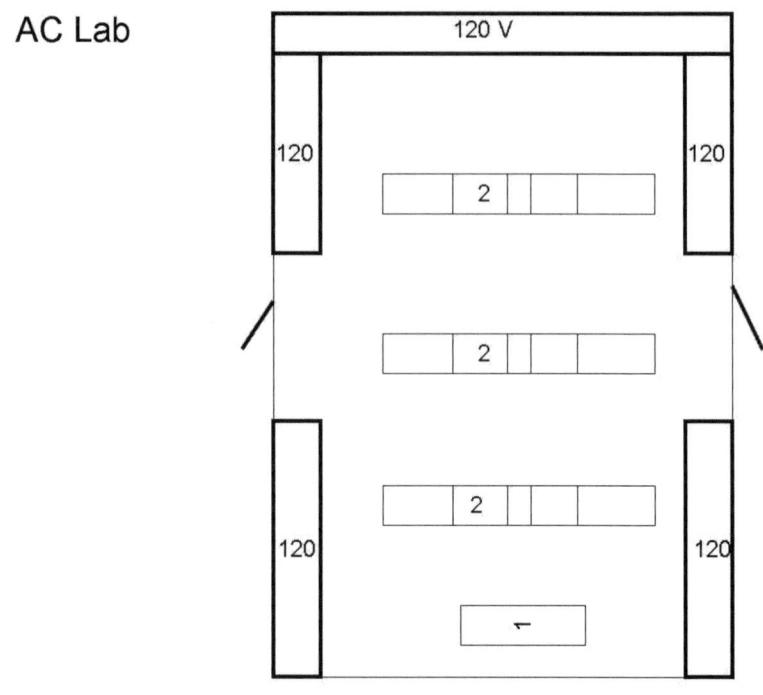

Figure 8-5

ic circuit, and then watch it operate. The trainer would also be able to insert various items (including faults) into the circuit to give the trainees troubleshooting experience.

Figure 8-5 shows the layout for the AC electrical lab. The table configuration was shown with the instructor table being labeled "1" and the tables that the trainees would set at arranged in three rows; these are labeled "2." The lab components for the electrical lab were raised around the perimeter of the classroom. This left a total of six panels for the students. The panels were very similar to those in the hydraulic and pneumatics lab in that all of the connections were front mounted. The trainees could then quickly wire up almost any circuit that was being discussed in class. The tables in the class were also very heavy duty. Thus, the trainees could use some smaller control boards and work at their desks. This arrangement allowed the trainers a tremendous amount of flexibility in how they would deliver their material.

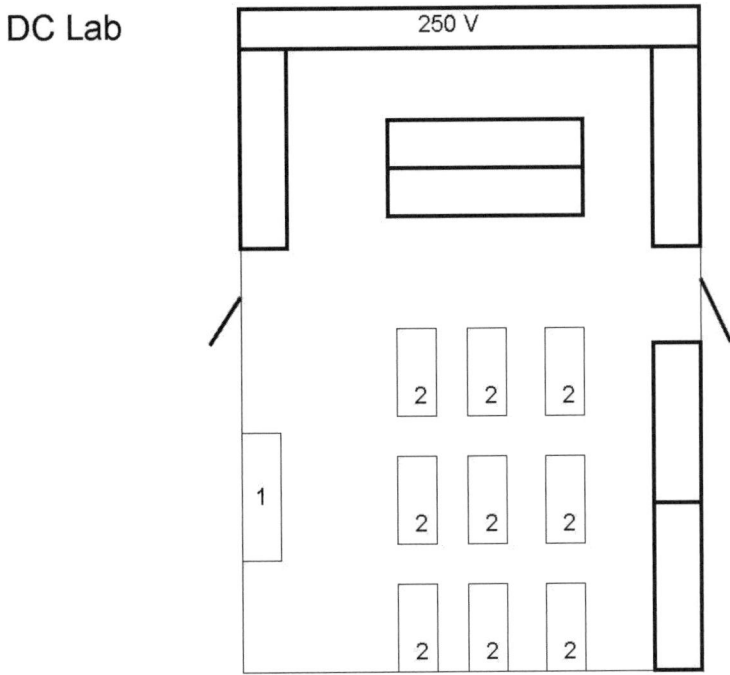

Figure 8-6

The DC lab is illustrated in Figure 8-6. This lab was quite extensive in that it actually had crane hoist motor control boards. The trainees could wire up and control circuits for hoists, trolleys, and bridges for overhead cranes. The lab also had stations where motor generator sets could be wired up and tested. The instructor's table, which is labeled "1" in the figure, was in front of the class. In turn, the class sat at the tables that are labeled "2." The boxes behind the classroom tables were actually interchangeable panels. At those times when 440-V AC classes need to be taught, the DC panels were traded off for the AC panels; the class was then taught in this room. Extreme care was taken to ensure that all the trainees worked very safely with this high voltage.

Computer-based training was offered occasionally in addition to the apprentice training program. Relevant topics were usually directed towards journeyman training, or some specialized equipment training. This training was typically offered by a computer simulator or an elec-

Computer-Based Training Classrooms

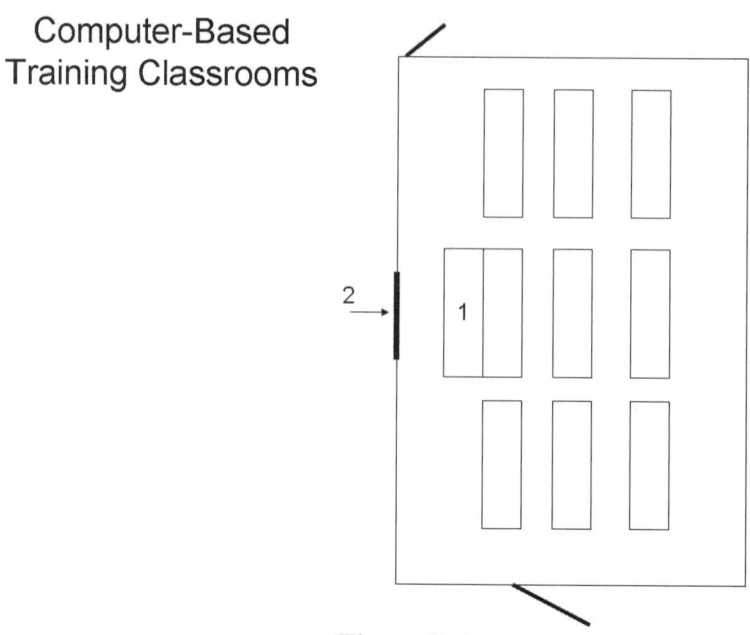

Figure 8-7

tronic computer-based training program. Classrooms for this training were set up with two trainees and computers per desk (see Figure 8-7). In many cases, the instructor controlled the computers in the room for part of the lesson and then let the trainees take control of the computer and finish the experiment or testing.

Many computer-based training programs are available today. When these programs are used in a classroom-type setting, the setup in Figure 8-7 proves to be very effective. There are also computer-based training programs that are available in stand-alone mode or on the Internet. In most cases, a trainee will take these types of classes on their own personal computer. Nevertheless, there is still a need for the classroom. Instructors are still useful for questions and answers, even if they provided their answers by e-mail or Instant Messenger.

The examples given in this chapter are based around one apprentice training program. However, the guidelines and principles outlined can be very effective in other settings. This material should be given close attention to ensure that maximum benefit is derived from the training program. Many other programs have been modeled very similarly to this program. All of these types of programs have been very successful.

9

ON-THE-JOB TRAINING

On-the-job training, or OJT, is one of the oldest forms of training. At its most fundamental level, OJT can be defined as two people working closely together so that one person can learn from the other. Whether the person teaching is called a trainer, a mentor, or a lead, the function is the same — to teach the trainees so that they can correctly and safely perform a task. OJT's strengths are in its flexibility and portability, all the while remaining an informal and human form of training.

There is no company, factory, or home business in the world where one person — the so-called expert (or SME) — has not helped a fellow employee learn a new skill. Everyone has a person definition of what OJT means. Essentially, it is a just-in-time training delivery system that dispenses training to employees when they need it. An OJT system can be as small or as large as the needs and resources of an organization allow.

Whether they know it or not, all companies have OJT systems in place. There is always someone standing next to another employee who says the magic OJT words, "Let me show you how to do that," and then teaches the other how to run or repair a machine or perform a task. This help is called unstructured OJT because it occurs haphazardly — the trainers teach the tasks as they know and remember them. Because of time or other pressures, important steps may be forgotten or simply skipped. As an unstructured system, no criteria are set for the quality of training nor are any of the records for the training maintained.

Building a Structured OJT System

In response to quality, ISO 9000, and budget constraints, companies have been required to organize their OJT process. Specific employees are designated OJT trainers, checklists of required skills (related to job responsibilities) are used to ensure that all employees receive the same training, and the training effort is tracked and recorded. Because the organizational structure explicitly supports the training, the process is called structured OJT. Structured OJT is more efficient than unstructured OJT and, in some reports companies develop a 60–80% decrease in training time.

How does OJT fit into a technical apprentice training program? We can refer back to our case study in Chapter 8 for the answer to that. In this apprentice program, at the end of each grade period, a week was set aside for OJT (it was also called in-department training). The OJT training was always linked to the classroom training that was provided during the grade period. Training coordinators from the training department were required to communicate with OJT trainers each grade period. The training coordinators were required to provide the OJT trainers with a list of topics that each student had completed for the grade. OJT trainers would then select a typical job in the department that would utilize the information and skills that the trainees had learned during the grade period. In some cases, the job would be simulated; but in other cases, the OJT may have involved performing a currently existing work order.

This effort was a challenge for OJT instructors because, in many cases, they had multiple trainees during the week. Although it was ideal to find an active job, in some cases, the training ended up being a demonstration or lab exercise.

Once the training week was completed, in-department instructors were required to submit information to the training coordinators as to what was accomplished during the week. This report would typically include the trainees' information, the job that was reviewed or trained, and the instructor's observation as to how the trainees were able to perform. The training coordinators were required to review all of the OJT forms. If a skills deficiency was noted consistently across multiple departments, the lesson plans for the grade period were reviewed. If there was a deficiency in how the material was prepared or presented to the trainees, trainers were required to improve the materials before the next class was taught.

In addition to this one week of OJT every 13 weeks, another approach was also used. In most maintenance departments, the maintenance technicians worked in pairs; therefore, it was possible to assign an apprentice to a skilled journeyman. This pairing was typically very effective when an apprentice was assigned to a highly skilled journeyman. Maintenance supervisors would periodically check with the journeymen to see how their apprentices were performing on the job. Although this training was not structured, it did allow the apprentices the opportunity to learn their trade from highly skilled journeymen.

This approach was not always successful. In many cases, due to workload constraints, apprentices were not always assigned to highly skilled journeymen. In this case, the knowledge transfer was somewhat limited. The approach also required discipline on the part of the maintenance supervisors to monitor the apprentices and journeymen constantly. If there were personality conflicts, they were to reassign the apprentices to different journeymen. They would also take this step if they noticed the apprentices were not developing good job skills.

Many improvements could have been implemented in this approach, particularly increased documentation; however, it was a useful approach to increasing the apprentices' job skills.

OJT Conflicts

Most companies want to implement a structured OJT system, but few achieve this goal. What typically happens is that the already existing, but unstructured OJT system begins to fail. In unstructured OJT, all the training time is casual, untracked, and therefore invisible. It never appears on a budget or time sheet. Structured OJT, on the other hand, makes this training time visible; at many companies, that visibility is risky. Once training time becomes real, supervisors often move to eliminate or convert it into production time, because they are, of course, measured by production output. Thus, many organizations destroy an existing unstructured OJT system when moving into a structured one.

In the training case study from Chapter 8, the 33 departmental maintenance supervisors were provided the guidelines for the OJT. They were expected to allocate the appropriate resources during the OJT week in the department to allow it to be effective. However, the in-department trainers were typically the best journeymen. Therefore, it was always tempting for the maintenance supervisors to send the trainers and the apprentices to work on a "hot" job to get it done. This assignment was not supposed to happen and, if it did, the OJT instructors were to notify the training coordinators, who would then include such incidents in their reports to the Maintenance Superintendent for the grade period. In turn, the Maintenance Superintendent would take appropriate measures to insure the disruption to the OJT did not reoccur.

Before making the transition (from unstructured to structured), beware of the fact that structured OJT only exists with the assistance, support, and understanding of management and production supervisors.

Training takes time. If supervisors do not allow enough time for preparation and training, they will thwart any structured OJT effort. If trainers cannot gain internal support from the organization's managers and supervisors, don't waste time trying to implement structured OJT. Instead work to strengthen the existing unstructured training process.

Similar to quality programs and thorough ISO-9000 programs, a structured OJT system will change the maintenance and production operation. Training assumes new importance — trainers are included when developing new production processes and are given time to carry out proper training. As with any system, trainers still need to be flexible when dealing with production and considerable customer needs.

When to Use OJT

The use of OJT as a trading method is determined by the following:
• Safety considerations
• Size of the training unit
• Geographical distribution of the trainees

When safety is a major issue, some companies use simulators or spare machinery for OJT training. The simulators are expensive; however, they allow trainees to experience potentially catastrophic circumstances without any danger to themselves or the facility. If trainers are going to use simulators or spare machinery for OJT training, they should use scripted scenarios. This will insure the training is useful, realistic, and consistent. In some companies, regulatory statutes require the use of simulators within the boundaries of the very structured OJT system.

If group interaction, interpersonal skills, or other personal communication objectives are part of the training, classroom instruction is probably the best delivery medium. The OJT instruction cannot fulfill interpersonal skills objectives because, by definition, OJT is one-on-one training. If the trainers have the requisite skills in specific training techniques, and there are sufficient controls in place over the training methods, OJT can be used for conceptual training in areas such as quality and customer satisfaction. In this scenario, trainers act as role models and instruct trainees through skills and behavior training. If a large number of employees need to be trained or the potential trainees are geographically dispersed, other methods of training (such as lectures or classroom settings) should be used.

OJT has one major drawback. It assumes the trainees are capable of learning and have the background skills, such as math and reading, necessary to perform the task. If prerequisite skills are not present, the OJT training will fail. OJT Trainers should be educated to recognize these deficits and respond appropriately. Occasionally, however, trainees will slip through even the most arduous pre-qualification exam.

From Unstructured to Structured OJT

Building on the example of unstructured OJT — where two people work together with one person formally teaching the other — a more structured system can be developed.

The next phase in the growth of an OJT system is to train the trainers various methods and techniques. Most unstructured OJT trainers have never been shown how to adapt to various student learning styles or how to most present materials effectively. This is one of the most critical and important steps in the transition from unstructured to structured OJT within a company. Untrained trainers will most surely cause an OJT system to fail.

Until now, the discussion has focused on casual trainers, that is, employees who work in a maintenance or production area and train when they have some spare time. As the structured OJT system continues to develop, some companies will increase their headcount, and assign full-time trainers to the OJT system. These trainers may have the responsibility to train new personnel, retrain existing employees who are changing jobs, or recertify staff on their current jobs. Full-time trainers do not need to be experts in all tasks. They may delegate some training task to part-time trainers who are experts in specific skill areas, but it is the full-time trainers' responsibility to ensure that all qualification checklists are completed. At this time, the company may assign employees as OJT system coordinators to track and record employee training progress.

Components of an OJT System

An OJT system has a number of components. Every OJT system contains all of them, but the degree of development of each varies significantly depending on the type of implementation.

Management Support

Management controls all the resources that are required; without their support, the OJT program will have little chance of success. These resources include attention to the details of OJT training, review of tracking reports, support with maintenance and production managers, and specific allowances for the budget, both in terms of dollars and time. The budget is used for:

- OJT training materials
- Computers
- Software for tracking students
- Training for the OJT developers and trainers
- Time for the training
- Time to work on processes and procedures

Figure 9-1 shows a pie chart with a sample breakdown of costs. This type of chart is useful in presenting the cost breakdown to senior management for their approval.

It is important to note that management may approve these expenditures; however, in the crush of maintenance and production priorities, they may never actually get a chance to spend the money.

Unfortunately, maintenance and production supervisors in any type of a company from financial to manufacturing are caught in the middle between two completely conflicting goals — training and production. If

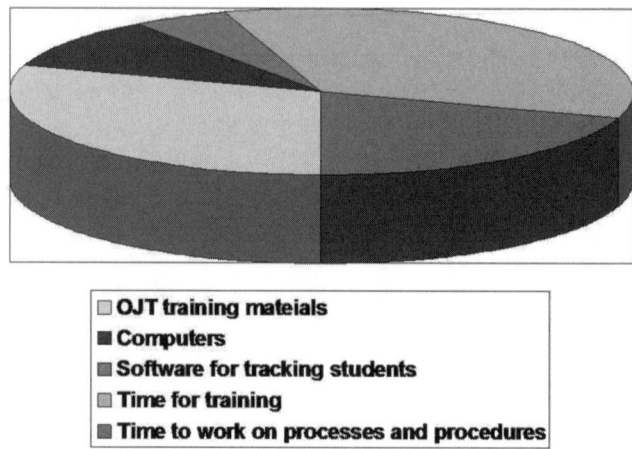

Figure 9-1 Cost Breakdown for OJT

supervisors take people out of maintenance or production and allow them to go into training, they lose production time. As a result, their short-term production numbers may be lower than those of other supervisors who do not allow for training time. Supervisors and line managers generally do not want to hear "Let me have your employees today, and I will give them back to you trained tomorrow." To supervisors, that means a loss of production today and they may not care about tomorrow.

Formal Trainers Support Process

Supervisors are very busy people who are not always known for their problem-solving or people skills. Generally, first-line supervision is the weakest link in the management chain. Trainers need another person, outside their organization, to go to for support, help, and advice on sensitive personnel issues.

Successful systems appoint an OJT coordinator from a neutral organization — usually the training department — who regularly interacts with the trainers. The trainers then work for the production area, but have a dotted line relationship with the support organization. In some systems, the trainers are transferred to the support organization and have a dotted line relationship with the maintenance or production organization.

Trainer protection is needed when trainers do not think employees are ready to be checked off on a task, but the supervisors insist that they be allowed to work and asked the trainers to check them off anyway. In this case, where do the trainers go for help and support? In other cases, when the trainers and the students have a conflict or when the trainers are asked to fix some non-training program, who supports the trainers? In extreme cases, the trainees may actually verbally attack the trainers and accuse them of not being good trainers. Safeguards need to be in place to protect the trainers.

Checklists

To succeed over the long term, structured OJT must be linked to other plant or company systems. Usually OJT is linked to ISO-9000, pay-for-performance, procedure sign off, or regulatory or mandated training systems. Linking the OJT system to another company system simply means that the OJT system cannot go away or be de-emphasized without also de-emphasizing the link system. It provides some protection to the funding, time available, and resource availability for the OJT system and its trainers.

The basis of a linkage to any system is the checklist, which proves that the students are capable of performing a specific task at a specific competency level. OJT systems without the checklist cannot be linked to others, and generally are not acceptable as a basis for ISO-9000 registration.

The checklist is the foundation of any OJT system. It lists the tasks that need to be trained, along with the administrative information such as the student name, training dates, and completion dates. It also has room for the signature of the trainer, the student, and the supervisor when the entire checklist is completed. A sample checklist is shown in Figure 9-2.

A skills checklist also adds structure to the process. With a list of specific skills that need to be taught, fewer tasks slip through the cracks — the employee is checked off as each task is successfully performed. Although checklists decrease the variation in skills being taught, some variation will still exist. During training, some trainers will let the students operate the equipment, whereas others will push the student to the side and show the student how to operate the equipment. Some trainers let students explore and make mistakes, whereas others make the students follow their exact directions step by step.

In some companies, the checklists and other OJT documentation are controlled and can not be changed without approval. In good OJT systems, trainers are on the signoff list for maintenance and production procedure changes; therefore, they know about them far in advance. This information gives trainers the chance to contribute to the decision-making process and ensure that the procedures are realistic. It also gives them the opportunity to update training materials and evaluations before the changes implemented.

In some cases of mandated training (ISO-9000 or other regulatory requirements) or in a pay-for-performance system, the checklist assumes greater importance. In these cases, the checklist may become a legal document, in connection to either an accident or to pay rates. Trainers should never take shortcuts or check off students until the students can perform the task to specifications. Otherwise, the trainers could be held responsible for any accidents or pay disputes.

OJT Training Materials

OJT trainers should not train off the top of their heads. This is dangerous and invites mistakes. In addition to the checklist, the trainees should receive procedures or other job aids. If the trainees can take notes

OJT Checklist

Week Ending _____   Training Week # _____

- Trainee Name

  _____

  Work Order or Preplan
  Number _____
  PM Number _____
  Short Description

  _____

  _____

  Objective: _____

  _____

- Check off each Job step as it is completed correctly
- Job Step 1 ___
- Job Step 2 ___
- Job Step 3 ___
- Job Step 4 ___
- Job Step 5 ___
- Job Step 6 ___
- Job Step 7 ___
- Job Step 8 ___
- Job step 9 ___

Instructor's Comments _____

_____

_____

Instructor Signature _____

Trainee Signature _____   Supervisor Signature _____

Figure 9-2

during training, they should be provided with a notebook or other organized method for writing. Trainees should not take notes on little pieces of paper; these will quickly get lost and the effort will be wasted.

If the trainees cannot take notes or get copies of either the procedures or equipment manuals, the procedures and manuals should be posted near the machine or work area. Many companies keep copies of procedures and equipment manuals in the supervisors' office, but rarely does anyone use them.

Some companies support trainers by giving them a room complete with computer equipment and printers, where they can go to develop the handouts or procedures necessary for the training. The time, equipment, and resources necessary to develop OJT materials are also vital to the trainers. One training scenario might go like this: "Here is a good component," say the trainers as they pick up a component from the line. "Notice how the edges meet and the color is even throughout the component. If the color is not uniform or the edges do not meet, reject the component."

Is this an example of a good form of quality training? Is it sufficient — will the trainees learn how to inspect the product properly? The answer is no to all these questions. The trainees do not know what is meant by "uniform color or edges meet." Do they have to meet exactly without absolutely any overlap or is a little overlap, okay? The trainees have not been told these criteria nor do they have a context within which to learn them.

Tracking and Report Generation

It is important that any OJT system be able to track and report on the training activities. Generally, these reports are statistical, providing numerical data on:

• Number of students trained
• Checklists completed successfully
• Number of training hours
• Percentage of successful students
• Cost of training — total and per student
• Return on investment of training

Many computerized training systems report on training development activities, testing results, and trainer effectiveness. Some of them also report on certification expiration and notify employees when they need retraining. Training administrators or coordinators who are responsible for the tracking system should also be prepared to produce reports supporting ISO-9000 audits, GMP audits, or annual budget reviews.

When checklists are created, the trainers should decide if the checklist is valid forever or carries an expiration date. When trainees complete the checklist, these expiration dates should be recorded and tracked within the system. In some cases, regulatory requirements will define retraining or recertification time intervals. Both trainees and trainers should be automatically notified when recertification training is required. Some computerized systems distribute critique sheets, so trainees can evaluate their instructors. The trainee critiques are no more valid in this environment than in classroom training environments; however, they can be useful for detecting trends and highlighting areas for improvement or praise.

Implementation

There are several techniques and procedures that should be considered before and during the OJT implementation process. First, OJT systems change the maintenance and production environment. They often require supervisors, management, and workers to change how they train and, in some cases, the way supervisors supervise. In hierarchically authoritarian organizations, OJT systems often intrude on some else's turf and this can be risky, to both the OJT system and individual trainers' career aspirations.

OJT Training Procedures

The OJT procedures call out the functions, responsibilities, and interfaces of the OJT system. Communications and interfaced responsibilities exist among the trainers, the maintenance and production requirements, and the supervisors. These responsibilities and communications are sources of potential conflict and misunderstanding. A complete procedure will govern the activities of the trainers and the supervisors; it will also describe the process for solving misunderstandings.

Specific items that should be covered in the procedures include:

1. Escalation procedure

What happens if the trainees fail to properly perform — how many times will they be trained and what happens if the trainee asked for another trainer?

2. Trainer Responsibilities

What specifically are the trainers' responsibilities with regard to procedures, development, and training?

3. Supervisor Responsibilities

What are the supervisors' specific responsibilities regarding signing training check list, releasing people from the line or their job responsibilities for training, and so forth?

4. Checklist Signoff

Should the supervisor conduct the final evaluation before checklist signoff or work with the trainer to conduct the evaluation? On what basis should the trainee be evaluated? Remember that the checklist could become a legal document for pay or regulatory purposes.

5. Organizational Responsibility

What are the organizational requirements to support the trainers and the training process properly?

6. Process and Equipment Changes

Are the trainers added to the signoff list for the equipment changes and revisions to production processes? How is this process implemented and maintained?

7. Conflict Resolution

In the event of a conflict between training and requirements for maintenance and production, what procedures are followed to resolve it?

Within any organization there are many other interfaces to be negotiated and resources to be assigned. Without clear procedures, turf issues will arise; generally, trainers end up yielding. It is best for everyone concerned to sit down during the development of the OJT system and develop appropriate procedures.

This chapter does not completely list the techniques used to implement OJT systems, neither does it fully discuss the ways in which an organization can be changed. In order to cover all of that information, several books on organizational development will be required. Trainers must seriously consider the effort required to implement the OJT program and the support they will really receive.

If realistically they will not receive any management or supervisory support, they should focus on developing a very low-level OJT system, keeping a low profile, and trying not to disturb the supervisors, until some results are being clearly shown. If trainers really need help (assuming they are the ones developing the OJT program), they should find assistance within the company, at the corporate level, or go outside for a good OJT consultant.

Be able to answer the following questions when making an OJT decision:

1. What is the motivation? Is it your idea? Do you have the political and financial influence to implement the system by yourself? Is this your assignment? If it is, can you negotiate a realistic level of effort to make the system successful?

2. What is in it for them? What will the rest of the trainees and supervisors in the plant get for their efforts? People will not buy into systems without first asking what is in it for them and getting meaningful answers. An attempt to implement any training system by edict may receive lip service, but as soon as the champion of the system turns away, it will be ignored and disappear.

3. What problems does the system solve? What will the trainees and supervisors be able to do better after the OJT system is implemented? There should be a list of specific problems, and the ways the OJT will solve them. Training should not be listed — everyone wants training to be better, but very few will expend the effort and resources to do it. Training must also solve a particular maintenance or production problem or help the organization achieve a specific goal. Problems to be solved should be listed specifically and solutions should be directly connected to or supported by the OJT system. This forms a foundation for the "what's in it for them?" needed to implement and support the training system.

After these questions are answered, an organization is ready to implement an OJT system. Successful techniques for bringing out the best of the company during implementation include:

1. Connect to Another Project

Trainers usually cannot lose by connecting OJT with another project, especially if that project has some urgency about it. These programs may include ISO-9000, regulatory programs, equipment breakdowns, quality, and customer satisfaction areas. The upside of this association is that the OJT program is likely to receive more support and resources than if OJT program is implemented alone. The downside is that the OJT program will have to share resources and authority; it may also lose some implementation priority. This usually turns out to be a fair trade.

2. Build on a Crisis

Every company has annual or semiannual crises. These crises usually revolve around missing production goals, late shipment, poor quality, or similar emergencies. If it makes sense, offer OJT as a partial solution. However, this solution has two downsides. First, the OJT sponsor may end up as a scapegoat if the problem occurs again. Second, after the emergency is over, all of the OJT support may disappear. The advantage of offering OJT as a partial solution is that the OJT program — at least for a limited time — receives resources that may allow it to build a foundation for the system and keep it operating after the emergency is over.

3. Start Small and Build

Many successful OJT systems begin like this: The OJT sponsor recognizes the current support level, designs the appropriate system at that level, and produces a system that is useful to the company. Supervisors slowly buy in as they see the usefulness of the system. Problems such as additional staffing are answered and, over time, the system grows slowly, receives more support, and eventually becomes part of the corporate culture. This approach does not work in all situations, especially when a fully-implemented system is needed immediately. It may not work in situations where another OJT implementation has failed.

4. Wait for the Right Time

If the OJT sponsors or implementers, after reading everything up to this point, feel overwhelmed, they are probably realizing they do not have much support, and also they have lots of responsibilities besides setting up an OJT program. They should maintain the status quo, continue to talk about the OJT program with supervisors and management, find people who will support the OJT system, and, most important, wait for the right time. Remember that OJT implementations have components of both training and organizational development. Successful implementers should have experience in both of these skills.

Finally, implementing an OJT system is a major undertaking; it may result in significant changes in a facility or factory. The implementers should not scrap the existing informal OJT system and insist on starting over. Instead, they should go with the current OJT flow, investigate and improve, and begin to control the informal OJT system. They should then harness the energy of OJT to help build better products (faster, better, and cheaper), customer satisfaction, and a better OJT system that will continue to develop, adapt, and provide better training as time goes on.

OJT Myths

Management and supervisors find many basic OJT concepts confusing or hard to understand. OJT implementers should understand and work through these misconceptions before they implement an OJT system. Otherwise, there will be trouble and confusion later on. These myths are listed in Figure 9-3 and explained below.

1. OJT is Free.

OJT systems take time, money, people, and energy to implement. When finished, OJT may be as expensive as classroom training, or produce a much higher return on investment for specific skills training.

OJT Myths

- OJT is free
- Training time is production time
- OJT is simply part of the job
- Anyone can be an OJT trainer
- Once implemented, OJT is forever
- OJT changes current organizational structures

Figure 9-3

2. Training Time is Production Time.

Maintenance and production personnel who are involved in a structured OJT program cannot produce at full capacity and trainers cannot perform in both positions at the same time. There will need to be extra personnel available to make OJT successful.

3. OJT is Simply Part of the Job.

OJT training is work and necessitates having trainers who agreed to perform the training and tracking activities. Employees should not be volunteered to become OJT trainers; they must want to be trainers. Documented OJT structures such as training procedures, list of task, assigned personnel, training materials, and equipment are essential.

4. Anyone Can Be an OJT Trainer.

Trainers should be selected carefully and then schooled in OJT training techniques and the use of OJT materials. Good OJT is not intuitive.

5. Once Implemented, OJT is Forever.

OJT systems require continuous review of OJT procedures and checklists, OJT trainer decisions, return on investment, resource allocation, and so forth.

6. OJT Changes Current Organizational Structures.

OJT changes the organization. It increases communication and forces power-sharing. But it is implemented on top of existing systems, not in place of them.

OJT is a necessary part of any technical training program. Properly developed, OJT reinforces any technical training developed and taught in a classroom or lab setting. Using the principles of OJT discussed in this chapter should improve its effectiveness in any company.

10

Measuring the Results

How is training justified? Some companies that have initiated specialized training for employees have tracked the results from a financial perspective. For example, one company saved over $4.5 million by training its maintenance technicians in the proper installation and maintenance of bearings. Another company saw its maintenance costs for forklifts drop 20% after training the workforce in how to operate the forklifts properly. Each manager should carefully reflect on how much damage is done to plant equipment because the operational and maintenance technicians are not properly trained.

Calculating ROI (Return on Investment) for Training

Calculating return on investment for training does not need to be a complicated process. An outline of how to calculate ROI for training is highlighted in Figure 10-1. The first step is to identify the losses that are currently being incurred. These losses can include lost employee productivity, excessive downtime on production equipment, or even high energy consumption. While investigating these losses, it is important to identify their root causes. In some cases, the losses can be created by organizational barriers or ineffective, work management processes.

Only when a skills deficiency is noted as a root cause for a loss should a training program be developed. This rationale excludes apprentice training development, because companies should always have the goal of moving apprentices to the journeyman skill level. There are assumed benefits for having a highly trained workforce, such as higher work productivity. Instead, what we are examining here with ROI assumes a certain level of worker skill already in place. This analysis will identify where skills are deficient, creating a loss.

157

Training ROI

- Identify the Losses
- Identify the Skill Deficiency
- Identify the Instructional Objective
  - To eliminate the losses
- Develop and Conduct the Training
- Measure the Results
- Quantify the Benefits

Figure 10-1

Once a loss due to skill deficiency is identified, it is necessary to identify the training that will be necessary to create a skilled efficiency. For example, if a higher-than-normal bearing failure rate is identified through the current work order system, and improper lubrication practices is identified as the root cause, a training program should be developed to highlight proper lubrication practices.

In developing the training program, as was discussed in Chapter 5, it is necessary to prepare a proper an instructional objective for the "Proper Lubrication Practices" class. With this instructional objective in mind, the trainers properly identify and develop training materials, set up the class properly, and conduct the training. By evaluating the trainees, either through a written exam or an instructor-observed lab exam, the trainers can evaluate the effectiveness of the training program. Whichever method is chosen, trainees should be able to demonstrate increased proficiency in bearing lubrication.

To this point, it is easy for the trainers to point to the exam scores to show the benefit of the training. However, senior managers will want additional evidence. The training program is developed based on losses that were occurring; therefore, after the training program is conducted, the business benefit should be shown by a reduction in the losses.

Let's consider that the bearings are being changed 20 times per year. Let's also assume that we are going to reduce the bearing replacements in half. The financial benefit can be calculated in several ways. First, the bearing usage itself should be reduced. This reduction of usage can be tracked and trended in the organization's CMMS/EAM system. In turn, the reduction in failures can be quantified by calculating the reduced bearing usage cost and the reduced labor cost to change the bearings. If the bearings cost $50 each, and it takes $200 of labor to change each bearing, then the savings equals $2500. This amount can be credited to the training.

Second, and more important, the impact that the reduction in bearing replacements has on the production process needs to be considered. Suppose the equipment downtime is $5000 per hour. If the reduction in bearing replacements eliminates 10 hours of downtime (1 hour per change), the savings is $50,000.

The total savings that can be credited to training in this example is $52,500. If the training class costs $10,000, there is a five-to-one payback on the training; if it is annualized, the training's payback will be in less than three months. These types of training benefits need to be highlighted to senior executives. This approach is also important because in very competitive businesses today some form of return on investment needs to be calculated for every expenditure.

With these points in mind, it is beneficial for each manager to ask "How is my workforce doing in keeping pace with the changing equipment technologies?" An honest evaluation will be beneficial for all managers. What are some of the indicators to examine? The following indicators will help in the evaluation.

1. Dollars per Employee

This indicator examines the actual average training dollars being spent per employee per year. Many companies train a select group of employees such as upper managers. But how much is actually spent on the main workforce itself, where a large return on investment is waiting for most companies? I'm not trying to minimize the importance of management training. However, many companies need to shift their focus toward training the operating and maintenance technicians. Averages for this indicator range from $1,000 (for the lowest) to $1,500 (for the highest) per employee per year.

Total Training Dollars
Total Number of Employees

This indicator is derived by dividing the total training expenditures by the total number of employees. This ratio gives in dollars the revenue per employee spent on training. The indicator can be calculated on a monthly basis and trended over time to insure that proper attention is being given to the training needs of the organization.

Strengths
This indicator is useful for trending the training expenditures and insuring that the proper overall level of training is being funded.

Weaknesses
This indicator fails to address the issue of training needs. In other words, is the training that is being funded the right training for the needs of the workforce at the current time? If this indicator is utilized exclusively, then some organizations may realize a false sense of achievement.

2. Hours per Employee
This indicator examines the actual average training hours being allocated per employee per year. Many companies will train a group of employees in soft skills such as diversity and team building training, but how much time is allocated for their hard skills, technical training? It is not that soft skills training are being minimized or are less important — these skills are essential to improving organizational effectiveness. However, how much is being expended in pure technical training for the operating and maintenance technicians? Technical training is also essential if the workforce is to be effective operating and maintaining the company's high tech equipment. Studies have shown that technical training and the interpersonal training should be split evenly.

Total Technical Training Hours
Total Number of Employees

Total Interpersonal Training Hours
Total Number of Employees

These indicators can be derived by dividing the total training hours in each category by the total number of employees. This gives in hours the time per employee spent on each type of training. These indicators can be calculated on a monthly basis and trended over time to insure that proper attention is being given to the training needs of the organization.

Strengths

This indicator is useful for trending the training expenditures by type of training and insuring that the proper overall level of training is being funded.

Weaknesses

Although these two indicators are more effective than the first one, they still fail to address the issue of training needs. Again, is the training being funded the right training for the needs of the workforce at the current time? Also, if these indicators are utilized exclusively, then a false sense of satisfaction may be achieved by some organizations.

3. Grade Reading Level

This indicator is more confidential that the previous two. It examines the overall grade reading level for the plant and expresses it as an average. This information is disturbing to some plants the first time they conduct a reading level assessment. Informal surveys conducted with seminar attendees over several years indicate that the average reading level in most plants today averages about eighth grade. The ranges are from a low of a third-grade average to a high average of second-year college. Another disturbing statistic is the illiteracy level in some plants. It has been found that, in some plants, one-third of the maintenance workforce is functionally illiterate. It does little good to print out a work order, if maintenance technicians can't read it. At the other extreme, one computer chip manufacturer requires its maintenance technicians to have a four-year college degree. This observation is not given to suggest that all maintenance departments must set that requirement, but to note that the chip manufacturer sets a different standard.

There is no formula for this standard. It is derived through standard testing and averaging the plant totals.

Strengths

This indicator is useful for highlighting the company's standard when it comes to basic skills. If the results are low, then remedial training in basic skills is required.

Weaknesses

If the testing and results are not kept confidential, the technicians may view it as a way to humiliate individuals or to highlight those who the organization may want to replace. The goal should never be anything like this. Instead, the goal is to identify whether the plant has an overall problem in the area of reading skills. Furthermore, the goal is to identify the basic skills needs in the organization and what steps must be taken to insure those needs are addressed through workforce additional training.

4. National Test Averages

This indicator examines the actual skill levels of individuals in the workforce through nationally recognized testing. The company can then evaluate how their employees match up with the averages in their own area or even in the areas where their competitors are located. A highly skilled and trained workforce allows a company to do many things that a company with marginally trained workers can not. For example, if a company's competitors have workforces with higher levels of skills, those companies may be able to move their workers into multi-skilling, operator involvement, or other high-performance initiatives.

This indicator has no formula. It is derived from national testing results.

Strengths

The indicator is useful for comparing work force skills from plant to plant within a company, or for geographical comparisons with a competitor's workforce.

Weaknesses

This indicator simply lists averages, which may vary. A highly-trained workforce may exist in one area, raising what would otherwise be a lower average score. This distribution can skew results when trying to compare areas. The averages are good to consider, but no comparison should be viewed as absolute.

5. Correspondence Training and Testing

This indicator examines the actual training scores of the plant employees in various technical courses. Managers can then identify and utilize the skills and talents of the various employees. If the training and testing are part of a pay-for-knowledge program or a pay-for-skills program tied to a needs–task analysis, then the scores can be used to promote individual technicians or to increase their pay based on applied skills.

There is no formula for this indicator. It is derived from the testing results.

As shown, this indicator will be the test scores themselves. The scores can be used to identify those individuals with high skill levels who can then be deployed in tasks that require the higher level skills and abilities. If the training and testing are tied to a job needs–task analysis, the results can be used to move individuals into a pay-for-applied skills and knowledge program, encouraging the technicians to constantly improve their skills.

Strengths

This indicator is useful for tracking the skills of individual employees.

Weaknesses

This indicator could be used to compare employees and rate them with their peers in an open forum. However, it should never be used that way. The testing and training progress of the employees should be kept at the highest confidence level. Any failure to provide security for this information can result in a complete lack of workforce support for any training or improvement program

6. Number of Trainers Compared to Maintenance Employees

This indicator examines the actual number of training employees per maintenance employee. This ratio will help an organization insure that the training programs for maintenance are staffed for success. Otherwise, the right amount of training will not be administered because the staff will not be available to deliver or coordinate the training.

$$\frac{\textit{Total Number of Trainers}}{\textit{Total Number of Maintenance Employees}}$$

This indicator is derived by dividing the total number of training employees by the total number of maintenance employees. This ratio highlights the staff support for the maintenance training effort. Depending on how much of the training is delivered by in-house instructors and how much is performed by contractor instructors, this indicator can vary dramatically. For example, the ratio may range from 1 training person for every 150 employees to as few as 1 training person for every 400 employees. In establishing a ratio goal for a plant, the training staff's workload should be closely monitored to insure effectiveness of the training.

Strengths
The indicator is useful for monitoring the level of staffing of the training department.

Weaknesses
This indicator has such a broad range, a company may feel that as long as its numbers are within that range, they are staffed satisfactorily. Without careful monitoring of the training workload, the training staff may find themselves overloaded. This overload condition will impact the overall effectiveness of the training and the company will fail to see the results from the training that should have been achieved.

7. OSHA Recordable Injuries per 200,000 Labor Hours

This indicator measures the number of OSHA recordable accidents per 200,000 labor hours worked. This indicator is a common one in the United States, but has little meaning in other countries because their monitoring organizations utilize a slightly different calculation. The formula itself is straightforward:

$$\frac{\textit{Number of OSHA Recordable Accidents}}{\textit{200,000 Labor Hours}}$$

The result is expressed as a ratio.

Strengths
This indicator is standard in the United States. All companies are required by law to track this information in the same format, allowing for an accurate comparison with other companies or industries.

Weakness
This indicator has no weakness.

Indicators Specific to Technical Training

The following indicators should be valuable to organizations attempting to cost justify training programs or trying to establish the value that training programs can provide to the overall profitability of the company. As the classic saying goes, "If you think training is expensive, try to calculate the cost of naïveté"

The following indicators will help a company calculate the cost of not training their employees, and further strengthen the business case for increased technical training.

8. Downtime Related to Operator Training

This indicator examines the actual equipment downtime that is caused by the operators' skill deficiencies. In many cases, extensive training programs do not exist for operators. Although regulatory programs such as Process Safety Management (PSM), OSHA regulations, and ISO-9000 certification require extensive operator training, few companies actually have documented standard operating procedures. If the maintenance and engineering department personnel were surveyed, what would be the equipment downtime caused by a lack of knowledge or skills on the part of the operations personnel? What if the cause could be recorded in the CMMS and added across a department or even the entire plant for a month? For a year? What amount of downtime would be identified?

Total Downtime Attributed to Operational Errors
Total Downtime

This indicator can be derived by dividing the total downtime attributed to errors the operators make that result in equipment downtime (in hours) by the total downtime (in hours). It has flexibility and can be derived for an area, a department, or an entire plant.

One of the most valuable alternatives to using the raw indicator is to calculate the cost of an hour of downtime to the company. It varies from type of equipment and type of process. The cost of downtime includes the cost of the lost product or throughput for the hour, not just the cost of idle labor or overhead for the hour. The argument that "there is no cost to downtime, we can make up the production" is analogous to the arguments companies use when they make rejects and rework items, but do not calculate the costs. Improvements in quality tend to come only after someone calculates what non-conformance or non-quality actually costs a company.

Estimates for per hour downtime costs range from $1,000 per hour for simple machining operations to over $40,000 per hour for line costs in a brewery, and over $100,000 per hour for downtime in a computer chip manufacturing plant. If the hours of downtime attributed to operational errors are multiplied by these figures, then training programs become easier to cost justify.

Strengths

This indicator is useful for tracking the hours of downtime caused by operational errors. However, it is even better utilized by tracking the reduction in downtime hours once a training program has been implemented. It is then easy to calculate the return on investment for the training.

Weaknesses

It is easy to use this indicator as a tool to identify operations personnel making mistakes and then punish them, rather than using the training programs to improve their performance. If the indicator is used as a tool for punishment, then the operations personnel will find ways to cover over the root cause of the problem and the losses will never be identified and eliminated. If this indicator is to be utilized, it must be with the proper goals.

9. Downtime Related To Maintenance Training

This indicator is similar to the previous one with the exception that it focuses on maintenance skill deficiencies. In many companies, apprentice programs for maintenance do not exist. Some companies have no structured training programs in place to progressively improve the current level of technical maintenance skills. Although regulatory programs such as Process Safety Management (PSM), OSHA regulations, EPA requirements, and ISO-9000 certification require extensive maintenance training and documentation, few companies actually have documented standard maintenance training and procedures. If the maintenance and engineering department records are surveyed, what is the equipment downtime caused by a lack of knowledge or skills on the part of the maintenance technicians? What if the cause could be recorded in the CMMS and added up across a department or even the entire plant for a month? For a year? What amount of downtime would be identified?

$$\frac{\textit{Total Downtime Attributed to Maintenance Errors}}{\textit{Total Downtime}}$$

This indicator can be derived by dividing the total downtime attributed to errors the maintenance technicians that resulted in equipment downtime or the lengthening of a repair period (in hours) by the total downtime (in hours). This indicator has flexibility and can be derived for an area, a department, or an entire plant.

Factoring the cost of downtime into the indicator can also provide the same benefits that it did in the operations example.

Strengths

This indicator is useful for tracking the hours of downtime caused by maintenance errors. However, it is even better utilized by tracking the reduction in downtime hours once a training program has been implemented. It is then easy to calculate the return on investment for the training.

Weaknesses

It is easy to use this indicator as a tool for identifying maintenance personnel making mistakes and punishing them rather than using the training programs to improve their performance. If the indicator is used as a tool for punishment, then the maintenance personnel will find ways to cover over the root cause of the problem. The losses will then never be identified and eliminated. If this indicator is to be utilized, it must be with the proper goals.

10. Lost Productivity Related To Maintenance Training

This indicator examines the productivity losses in maintenance activities caused by a lack of skills and knowledge in the maintenance work force. This indicator is more difficult to calculate than the previous two because the measure typically involves a subjective assessment by a supervisor or manager of the maintenance technician's actual work activities. The items highlighted here include any time that is lost because an individual does not have the skills or knowledge to perform the work in the most effective and efficient manner.

Estimated Lost Time Due to Lack of Knowledge or Skills
Total Time Worked

This indicator can be derived by dividing the estimated time and productivity lost due to a lack of knowledge or skills (by the maintenance technicians) by the total time worked. It can be tracked by type of work, specific job, specific skill, or any parameter that might be useful in identifying a potential training need. These needs can then be prioritized based on the amount of time lost. The wage rate and impact on the time to perform the task (including the downtime of the equipment) can all be calculated in dollars; the required training can then be cost justified.

As the training is conducted, the increased productivity can be tracked and the results expressed in dollars saved. This measure is effective in calculating return on investment for the training program.

Strengths

This indicator is useful for identifying training needs, justifying the costs of training, and calculating the return on investment once the training is conducted.

Weaknesses

This indicator requires a subjective opinion on the amount of lost productivity attributed to a lack of knowledge or skills. This opinion may be disputed by some in the organization. However, if similar work can be found in other departments and plants, and comparisons drawn from these examples, then some of the disputes can be eliminated.

The indicator can also be used to compare individuals. In itself, this is not wrong, but if one uses this data to reprimand or even criticize an employee, the use of the indicator will eventually be discontinued. Proper focus and understanding of the indicator's use is critical to its success.

11. Percentage of Maintenance Rework Related To Maintenance Training

This indicator examines the amount of rework that is performed by the maintenance workers. Rework is defined as work that, because it was not done completely and correctly the first time, must be adjusted, changed, or perhaps done completely over again. What amount of this type of work occurs because there was a lack of knowledge or skills on the part of the technician who performed the work the first time? This type of rework needs to be identified because it can be eliminated through a focused training program.

> ### *Maintenance Rework Due to Lack of Knowledge or Skills*
> ### *Total Maintenance Work*

This indicator can be derived by dividing the total hours of maintenance rework by the total hours of maintenance work. The resulting percentage highlights the opportunity for improvement by training the technicians so they can do the job right the first time. The losses in labor costs and material costs can be calculated for the rework and used as justification for the training program. In addition, the cost of the downtime incurred during the rework should also be factored in the savings potential. The return on the investment for the training can be calculated by trending the decrease in maintenance rework. The return on investment is calculated by the cost of the training program compared to the savings in reduced maintenance labor, materials, and plant equipment downtime.

Strengths

This indicator is useful for calculating the potential benefits of eliminating or reducing maintenance rework by increased maintenance training.

Weaknesses

This indicator's weakness is the slight chance that some of the maintenance rework will not be properly identified. In addition, the information may be misused as a way to punish technicians, instead of using it to focus the training effort.

12. Average Training versus Payroll

This indicator examines the actual average training dollars being spent compared to actual payroll. It is similar to the training dollars per employee calculation. However, this indicator compares the actual training cost as a percentage of the plant's payroll. Typically about 2.3% of the plant payroll should be spent on training. Again, the question must be asked: What type of training and for whom? Is it management training, interpersonal training, or technical training? The indicator is calculated as follows:

Total Training Dollars
Total Plant Payroll

This indicator can be derived by dividing the total training expenditures by the total plant payroll. The resulting percentage shows the amount of the plant payroll allocated to training.

Strengths
The indicator is useful for trending the training budget as the plant payroll increases or decreases. It insures that the proper level of training is budgeted.

Weaknesses
The indicator has no major weaknesses. It is essential to have plant management commit to the proper percentage. Then as the plant increases or decreases the number of employees, the proper level of training is assured.

Problems with Technical Training Programs

With the challenges mentioned at the start of this chapter, it is apparent that most training programs are experiencing problems, if not outright failure. The most common reasons for training program failures are discussed in this section and illustrated in Figure 10-2.

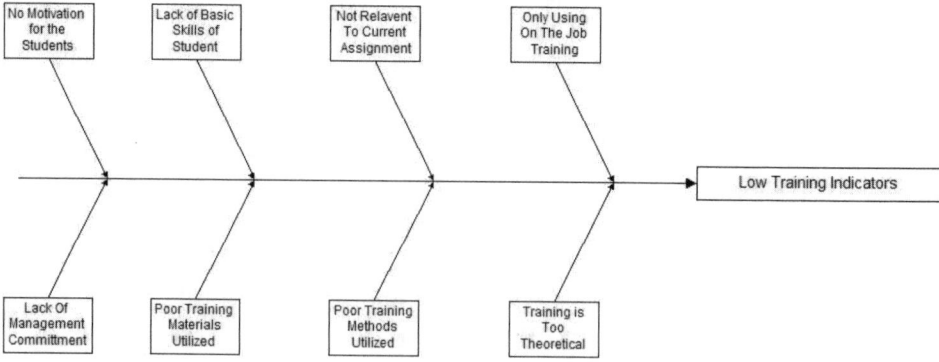

Figure 10-2

1. Only Using On-The-Job Training (OJT)

This problem is common in organizations without the resources to develop a structured and documented training program. The managers rely on the other technicians to show an operator how to run equipment or, for a maintenance person, how to maintain, troubleshoot, and repair the equipment. Because the employees giving the on-the-job instruction also have other jobs, they are hurried to convey the minimum of what the trainees need to perform the basics of the job.

The detailed instruction that makes a technician efficient and effective is never provided. The trainees then may make great personal efforts to perform the job, but are never efficient and effective. In addition to wasted labor resources, the equipment operation is affected and plant capacity suffers.

If companies are using only on-the-job training, they should remember that the method also teaches trainees someone else's bad habits. On-the-job training is never successful unless it is supplemented by additional training.

2. Training Is Too Theoretical

In some training programs, the information is taught directly out of textbooks and never supplemented with any actual application of the material to the technician's work assignment. Instead, the technician must make the application. Some will be able to meet this challenge, but most will not. This limitation impacts the effectiveness of the training; the return on investment in the form of improved technician performance will not be realized.

The solution is to make the training job relevant. This may involve a duty or task analysis of the job, and blending theory, hands-on lab activities, and on-the-job training to assure that the trainees will be able to master the new skills being taught. Without being able to make the training job relevant, there is little chance for retention on the part of the trainees.

3. Training Not Relevant to Current Assignment

Once training has taken place, it is important that the trainees put the new skill and knowledge to use. If they must go weeks or even months before they can use the training on the job, their retention rate falls to a very low level. In fact, if months have elapsed, they may need retraining.

The message here is short and concise: provide the right training at the right time and at the right level. Unless these criteria are met, the organization will waste its training expenditures.

4. Poor Training Methods Utilized

The issue underlying this problem is a lack of flexibility in the training delivery. A good training program uses a blend of materials and presentation styles. These include:

• Traditional classroom settings
• Computer-based training modules
• Video-based training modules
• Self-paced correspondence materials
• Satellite training broadcasts

If one style of presentation is used exclusively, then trainees become bored and uninterested. By using a blend of techniques, the instructors can insure the training program always holds the attention of the trainees.

5. Lack of Basic Skills of Students

Unfortunately, lack of preparation is becoming a large problem in many companies today. As individuals leave high school, they lack the basic skills in reading, writing, and mathematics. In turn, companies have no choice but to develop internal basic training programs designed to insure a minimum competency in the workforce. Although some companies do not have a problem in this area, due to acceptance testing, other companies do not have favorable geographic locations, compensation systems, or work environments to assure a pool of highly-qualified candidates from which to select employees.

6. Poor Training Materials Utilized

Poor training materials may indicate insufficient funding for the training program. Some training programs using photocopies of copyrighted materials. In other cases, materials are many years old are used and even recirculated among trainees. If a training program is to be effective, appropriate training materials must be provided. Many quality sources provide excellent training materials.

7. No Motivation for the Students

This area is sensitive in many companies. What is the incentive for an employee to want to learn new skills? Some issues include:

- Will there be a pay increase?
- Do the new skills lead to a new job?
- Is training required to perform the current job due to a technology upgrade of the equipment?
- Is the training designed just to insure the employability of the technician?

The motivation may be all, some, or even none of these reasons. Keep in mind: no matter what is used, something will have to motivate the employees to take the training and apply it. It is up to the management of each organization to find the right motivation for its workforce.

8. Lack of Management Commitment

One fact is clear: training is expensive, but ignorance is even more costly. In many cases, however, when cost reductions are made, cuts begin in training and maintenance. These cuts are really a problem in maintenance training. Management needs to be committed to training its employees if the company is to be competitive in the next decade.

Training needs to focus on the return on investment the company will receive. Therefore, training must focus on resolving identified issues. These issues will have to be rated by the financial impact they have on the company. They will have to have a training needs analysis performed to identify the training requirements. The training will have to be developed or purchased. The results of the training will have to be tracked. The improvements will have to be expressed in a dollar amount so that the return of the training investment can be calculated and the effectiveness of the training evaluated.

Unless these activities are undertaken in a detailed and structured format, management will never commit to a sustainable training program.

11

Continuous Improvement in Training

People do things for the strangest reasons. They also don't do things for the strangest reasons. To improve the results we get from training, we need to be able to understand why people do and why people don't do certain things. In this chapter, we will identify some of the roadblocks to training along with ways to overcome them if we're going to continue to make training successful.

Performance Discrepancies

To be successful in improving training programs, we need to be able to identify performance discrepancies — the difference between what current performance is and what future performance should be. These performance discrepancies are typically visible when unacceptable work practices and resulting wastes are identified. The unacceptable work practices can be any deviation from the roles, responsibilities, and processes flows that were developed in Volume 3 of the Maintenance Strategy Series. The personnel involved should have been properly trained to fulfill those roles and responsibilities. If training is to be improved, then the performance discrepancy must first be identified before any improvements can be made.

There is a disciplined process to identifying performance discrepancies and improving (or providing) the training necessary to eliminate the problem. This process, pictured in Figure 11-1, will be described in the remainder of this chapter.

**Analyzing Maintenance
Performance Problems**

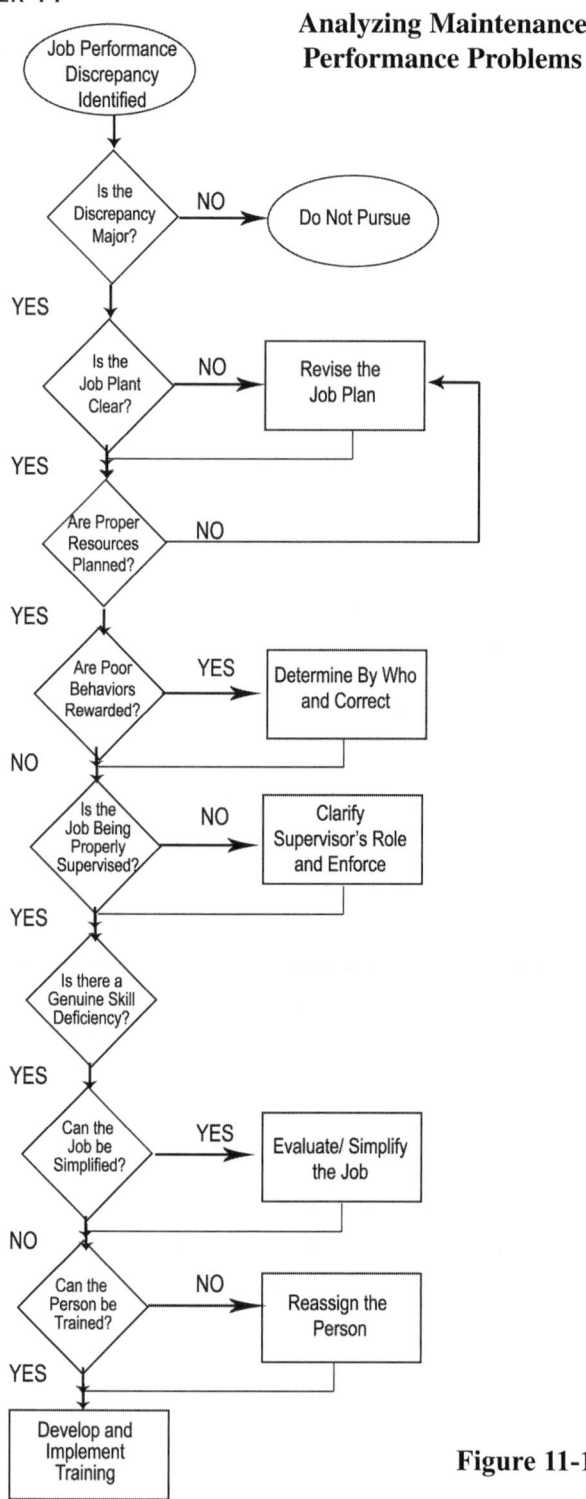

Figure 11-1

First, the group of people whose performance is said to be lacking must be identified. They could be maintenance journeymen, apprentices, or even operators. The individuals most likely to identify these performance discrepancies are the supervisors, maintenance planners, and maintenance engineers. Once identified, the perceived performance discrepancy needs to be described in the form of what is happening vs. what needs to be happening.

ROI and Training

Discovering a performance discrepancy does not automatically mean that someone rushes into developing a training solution to solve it. One of the first things we should ask is, "What happens if we ignored the performance discrepancy?" If the answer is "not much," then the performance discrepancy is minor and does not necessarily need to be addressed. The second question to be asked is, "What would happen if we changed behaviors to eliminate the performance discrepancy?" To answer this question, we would need to project what the results would be from changed behavior. These results can be described in terms of eliminating waste, increasing productivity, or increasing equipment throughput. With this value identified, we would calculate the cost of making the change. By comparing the benefit of the change and the cost of training, an estimated ROI (Return on Investment) can be calculated. If it's plain that the cost would be excessive, then again the discrepancy does not need to be addressed.

However, if the discrepancy is not trivial, and it is cost-effective to eliminate the discrepancy, then the following steps should be undertaken:

• Develop additional detail about the cost of the discrepancy.
• Find out the causes and the possible solutions and cost to implement.
• Calculate the detailed return on investment.

The return on investment will tell you whether it's time to do something about the performance discrepancy and what cost-effective solutions can be implemented. This information can then be presented to senior management to assure support for the project.

As was discussed in Chapter 10, calculating the return on investment for training can become complex. When working with a performance discrepancy, consider the following areas:

- Productivity — Do people waste time as a result of the performance discrepancy?
- Spare parts — Do people use more spare parts or stock or spare parts or create spare parts shortages due to the performance discrepancy?
- Equipment — Is there damage to the equipment that results from the performance discrepancy?
- Equipment efficiency — Does the performance discrepancy impact the efficiency of the equipment?
- Equipment availability — Does the performance discrepancy impact the availability of the equipment?
- Energy usage — Does the performance discrepancy increase energy consumption?
- Accidents — Does the performance discrepancy increase the frequency or the severity of potential accidents?
- Regulatory compliance — Does the performance discrepancy create or increase the likelihood of a regulatory violation?

Once the impact of the performance discrepancy is clearly understood, we should also calculate the frequency of the performance discrepancy. A performance discrepancy that causes a problem once a year might be ignored. However, the same performance discrepancy that causes a problem once a day would demand that a solution be found.

Thus, when deciding whether or not to implement a solution to a performance discrepancy, the costs to implement it must be measured against the costs to ignore the discrepancy. A solution should be implemented only when there is a definite cost benefit.

Developing an Action Plan

Once identified, a performance discrepancy demands action. Explore quick fixes before spending time and resources on further analysis. You may find all that is required is a quick and easy remedy, such as uncovering mistaken expectations, not providing proper resources, or not supplying proper feedback to the employee. Before beginning serious analysis of the performance discrepancy, check to see whether there are one or more obvious obstacles to the performance that can be readily eliminated or minimized. To understand if there are quick fixes to a problem, examine the following areas:

- Do the employees know what's expected of them? Have they been previously trained or otherwise told what's expected of them, or is it just assumed that everyone knows? If there are written standards, are they incomplete or unclear?
- Can the employees tell you what they're expected to do or accomplish?

From Figure 11-1, consider the following:

- Is the job plan clear? If it is not clear, then there may be mistaken expectations. The planners may intend one outcome from performing the work, supervisors may see another outcome, and the maintenance technicians assigned the work may even see a third.
- Is the job plan complete and detailed? If it is not and tools or parts are missing from the plan, then it is likely there will be a performance discrepancy in performing the planned work.
- Are proper resources planned? Coordination among the various crafts assigned to the work order, tools, equipment, contractors, etc., may be needed; otherwise, a performance discrepancy would exist.

Although the performance of the maintenance technicians would be the focus of the initial performance discrepancy, the examples above clearly indicate that the focus of any training program would not be for them, but rather for those involved in arranging the logistics for the assigned job. In many cases, what is identified at first as a performance discrepancy is actually a breakdown in the work processes.

Rewards and Performance

When faced with a performance discrepancy, another area to consider — and highlighted in Figure 11-1 — is that poor behaviors are being rewarded, thereby contributing to the performance discrepancy. For example, the planners and supervisors may plan and supervise a job to be executed properly. Meanwhile, operations may encourage the maintenance technicians to take "shortcuts" during the job so that the equipment can be put back into service faster and production can resume. Operations may even reward this behavior through extra gloves, coffee, doughnuts, etc.

If the desired performance is being discouraged, even subtly, corrections need to be made. Consider that in maintenance it is quite common to reward a hero, someone who can make reactive repairs very quickly.

True craftspeople, who take a little longer to make high quality repairs, are often spoken of negatively because they take longer to complete a job; however, their jobs never require re-work. Here are cases of rewarding the wrong behavior. To encourage proper behaviors, any negativity (particularly from supervisors or planners) must be eliminated or at least reduced. Individual managers will have to do all they can to create or strengthen positive or desired behaviors.

The cause of performance discrepancies may be the organization's pattern of providing hidden rewards for undesirable performance. If the analysis reveals that poor performance is being rewarded, the incentives for performing poorly must be removed. In the case of performing reactive work quickly, rewards often come from production supervisors whereas the behavior is often discouraged by maintenance supervisors. Here, it would be good for the maintenance and production supervisors to discuss what good maintenance behaviors actually encompass. With a clearer understanding, the production supervisors may stop encouraging poor behaviors.

In other situations, employees are neither rewarded nor punished for doing the job right or wrong. In fact, there are no consequences for doing the job at all. In these cases, policies and procedures should be put in place that encourage the proper behaviors and discourage the wrong behaviors (or not doing the job at all). If you want someone to perform in a particular manner, it is beneficial to reward what you value.

The Role of Supervisors

One additional area that needs to be considered when dealing with performance discrepancies is the supervisor's role. In many cases, maintenance supervisors are not being properly utilized. For example, a typical job description for supervisors requires them to be on the floor with their technicians for 6 out of the 8 hours during a shift. This schedule allows them time for proper mentoring, coaching, or on-the-job training. However, in many cases today, supervisors are off the floor in meetings and doing paperwork for 6 out of the 8 hours during a shift. What at first glance seems to be a performance discrepancy is actually a failure to properly observe the supervisor's roles and responsibilities. Before developing a training program to correct what appears to be a performance discrepancy, all secondary causes need to be investigated and corrected first.

Is Training Appropriate?

When a true performance discrepancy is detected, do not automatically assume that the problem is a training one. And do not assume that the solution involves teaching or training, even when the performance discrepancy involves a skill deficiency. Before taking action, determine whether the performance discrepancy is due to a genuine skill deficiency.

It is always important to decide whether a skill deficiency is due to some form of forgetting or the fact that the skill never existed. If the skill never existed, then training is likely to be indicated as the solution. But if the skill once existed, and is now lost or forgotten, training would be a more expensive remedy than is necessary. More likely, practice and feedback are all that is needed.

If employees could previously perform a job, another question to answer before developing a training program is, "Is the skill used often?"

Any time performance is something other than what is desired, and there's reason to believe that the desired performance could be within the employees' capabilities, check to see whether they are receiving regular information about the quality of their performance. When infrequently used skills slip, it is likely that a lack of feedback is a probable cause. Another probable cause to consider in such cases is simply a lack of practice. However, if the skills are used frequently but have deteriorated despite regular use, supervisors should provide periodic feedback in order to help maintain the required level of performance. If the skill is used infrequently, maintain the level of performance by providing a regular schedule of practice with feedback and observation and coaching by the supervisor.

Even when a genuine skill deficiency exists, any solution to the problem should be weighed against the possibility of simplifying the job by changing some aspect of the job, thereby overcoming the effect of the skill deficiency, at least in part. Opportunities to incorporate one or more kinds of performance aids — such as checklists, detailed job plans with step by step instructions, signs, labels, color coding, and similar job aids — can almost always be found.

For example, if a problem exists due to operators missing inspection points on an equipment inspection, then a detailed checklist could be developed that specifically details each inspection point, including acceptable and unacceptable readings. With this level of detail provided on the inspection sheet, it would be unlikely that any operator would miss an inspection point or misinterpret any readings. Using this detailed check sheet would eliminate the need for training the operators.

If training seems to be the only remedy, then on-the-job training with follow-up coaching may be easier and cheaper, and just as good as the formal training may provide.

Before settling on a formal training solution, you should always look for subtle impediments to performance. Look for factors that might be getting in the way of employee performing as desired, such as:

- Lack of authority, lack of time, or lack of tools
- Poorly placed or poorly labeled equipment
- Poor lighting and uncomfortable surroundings
- Lack of direct information about what to do and when to do it
- Competition from secondary tasks within the job itself
- Actions or inactions of other people who haven't been properly notified about the work being performed

There's usually no excuse for secret agendas or conflicting policies. Keep in mind that if maintenance technicians can achieve good performance but aren't, there is a reason. Seldom is that reason either a lack of interest or a lack of desire. Most people want to do a good job. When they don't, it is often because of an obstacle in the manufacturing or process world around them.

Be aware that the assumed cause and effect are not always accurate. The true origins of performance problems are not always obvious. Performance problems often surface downstream from the place where they originate.

Any time it seems that there is no cause for a performance discrepancy — that is, poor achievement doesn't seem to be due to a lack of skill or motivation — keep looking at other areas because you haven't yet found the obstacle. Widen your search.

Keep in mind: first you state the problem and the cause; then you devise solutions. Having a training solution in search of a problem is not a recommended approach.

From Figure 11-1, you should determine whether employees have the capacity to do the job required and whether they would fit the job mentally and motivationally, even if their performance was brought up to standard. If the answer to both questions is yes, go ahead with your analysis. If the answer is no, replace the performer with someone more suited to the job.

Summary

A company installed a high-speed packaging line and found they were having problems keeping the equipment running properly. Several contributing causes led to the excessive downtime. They included:

• Equipment jams
• Slow maintenance response times
• Maintenance technicians with incompatible skills making repairs

The solutions were more than just "do training." The jam problems were related to the operators not being able to clear the jams and restart the equipment without maintenance support. The solution to the jamming problem involved a slight re-design to the equipment to provide an opening for the operator to clear the jam. In addition, a very specific job procedure was written for the operator to clear the jam. The operators were then provided on-the-job coaching on how to follow the step-by-step procedure. The operators were then qualified to clear any jams safely and correctly.

The slow maintenance response times were not a training issue, but rather an organizational issue. The organization was using a centralized organization and needed to have an area maintenance organization instead. When the maintenance technicians were given area assignments, based close to the equipment, the response time decreased dramatically, with a resulting increase in uptime.

The problem with maintenance technicians with incompatible skills making repairs was actually a supervisory problem. The operations personnel were asking the first maintenance person they saw to make repairs. However, not all of the maintenance technicians were trained to make repairs on the new line. Therefore, it took some technicians longer to make the repairs. In some cases, the repairs were not correct and the machine failed again in a short time period. An agreement was made that the operators would only call the maintenance supervisor with equipment problems and the supervisors would dispatch the correct maintenance technician.

Still, there was a training component to the solution. When the line was purchased, a select few maintenance technicians were provided training on the new equipment. This small number limited the pool of technicians the supervisor could actually assign to make repairs on the equipment. In response to the problem, the decision was made to train additional technicians on how to maintain and repair the new equipment properly. This training started slowly, with only two technicians a month being trained by a factory representative. The course was one week in length and was provided to each journeyman technician in order of seniority. Once the training was completed, combined with the other changes, the line ran with a high level of uptime and actually was able to produce slightly above the design specifications.

The moral of the case study is that training by itself will not always solve a problem. However training is usually a component of any technical problem with a company's equipment.

In answering the question "What do I do now?" take the following steps:

- Collect all the potential solutions that address the issues revealed by the performance analysis.
- Determine or estimate the cost of implementing each solution.
- Choose the solution that will add the most value or solve the largest part of the problem for the least effort.
- Draft a brief action plan that describes, for each solution, what steps would be put into practice, and who will do the work.
- Execute the most cost-effective plan.

By following the guidelines that are provided in this chapter, non-training related problems can be identified and corrected, thereby focusing training budgets on those performance discrepancies that need training solutions.

12

Managing the Next Generation of Technical Employees

Author's Note: It is beyond the scope of this text to deal with the specifics of the difference among all generations, whether Baby Boomers, Generation X, Generation Y, Generation Z, etc. In some cases, the material will generalize when making the comparisons among them. My approach is not meant to minimize these differences, but rather to allow the reader to note the major gaps among the Baby Boomers and the subsequent generations that are following them. The main thrust of this information is to show that existing managers will have to adapt their management styles and policies if they are going to be successful in managing the new workforce as the Baby Boomers retire. In addition, the observations in this chapter are not intended to pass judgment on any of the qualities of any generation of workers; rather they are made to help managers realize and adjust to the differences of each generation of workers.

This chapter will examine the difference between the existing workforce, comprised of "Baby-Boomers" and the emerging workforce of Generation "X" and "Y" employees. Generation X is defined as the generation following the Baby Boomers. Generation Xers were born between 1965 and 1980. Generation Y includes individuals who were born from 1980 to 2000. This explanation is important because companies are or soon will be experience the transition from the Baby Boomers to the subsequent workforces. To complicate matters even more, more than 60% of employers say they are experiencing tension between employees from different generations. By evaluating some of the generation gap issues, management styles may be adjusted to reduce or eliminate this tension. In addition, the profiles of all three of these groups show that training will have to be adapted to be effective for each of the groups.

Baby Boomers

The baby boom generation is now reaching retirement age. Nearly 1/3 of all Americans — approximately 76,000,000 people — were born between 1946 and 1964. As boomers reach traditional retirement age, the question arises of how corporations will survive the anticipated mass exodus of skills, experience, customer relationships, and knowledge. These departures will become a critical brain drain. To survive, companies will have to adopt an integrated approach to manage the aging workforce based on the right managerial mindset, with a clear understanding of their business models in relation to the aging workforce. Companies will need a different leadership approach to be successful.

Although many baby boomers will be eligible to retire, the question arises "Will they?" Many baby boomers have seen their retirement pushed off into the future by the economic recession of 2008–2009. In addition, in 1950 a person age 60 may have been considered old, but today, this perception has shifted to age 70 or even older. Therefore, many baby boomers may choose not to retire until reaching a mandatory retirement age (in some cases 65+).

In 2008, the average worker in developed countries retired at age 62. These workers then expect 20 years or more of active life in retirement. In the next decade, a number of older employees (of retirement age) will simply discard the concept of retirement as it is known today; the concept will be replaced by a more flexible view of work. In this new workplace environment, work will be intermingled with periods of relaxation and travel. Already as many as 34% of all workers in the United States say they never plan to retire.

If baby boomers choose to remain in the workforce, what can be done to insure they have rewarding and productive job assignments before retirement? What type of training and job challenges can be presented to them to keep them motivated and engaged?

The challenges posed to companies by an aging workforce coupled with increasing employee turnover (especially among Generation X and Y employees) and the developing skill shortage are now beginning to gather greater management attention. The aging workforces need continued respect and an appropriate work environment. They especially need learning opportunities, which are often neglected by companies. The delayed retirement also means there will need to be a blending of Baby Boomers, Generation Xers, and Generation Yers.

This blending can be problematic. Research shows that more than 70% of older employees are dismissive of Gen X and Y's abilities. Meanwhile, nearly half of employers say that Gen X and Y's employees are dismissive of the abilities of their older co-workers. How can these various generations of workers co-exist in companies today? These workers will require a variety of diversity training programs focused on the contributions each generation brings to the company and how each is needed for the company to move successfully into the future.

What are some of the changes that can take place to allow the baby boomers to still contribute to the success of the company? For one, job sharing between older employees with similar skills should be implemented. This concept, already practiced in some companies, allows older employees part time work and income, interspersed with periods of no work. These periods of no work would allow the older employee time for relaxation, volunteer work in the community, or travel. Studies have shown that already 50% of the baby boomers would prefer this type of arrangement.

In addition, contract work for older employees is an option. This approach would allow companies to arrange their work in project-type formats, where the older employees work on a project, from start to finish and then take time off until another project that interests them becomes available. Keep in mind that project work can involve developing training materials or providing specific on-the-job training to younger employees.

These new arrangements will create new career paths, flexible work opportunities, different paths of personal learning, health as a core value, and customized employment. Why should companies do this? From their perspective, it provides a way to retain its store of critical knowledge while expanding its creativeness and productivity for continued competitiveness.

However, now comes the difficult part — blending the baby boomers with the next generations. This process begins with understanding the differences among the generations and how to bridge the gap. Yet one-third of Human Resource professionals say their companies are doing nothing to prepare for this shift in work force demographics. If companies are going to make the transition from the Baby Boomers to Generation X and Y, they will need to invest the same level of resources to optimize the relationships among the workforces as they have invested in optimizing their processes and technologies in the last decade.

Generation X and Y

The next generation of younger workers is in high demand. In 1980, over 50% the workforce was under 35. Now, that percentage has leveled off at about 38%. This shift means there will not be multiple younger workers applying for every opening. Competition for the best educated and most skilled of the Gen X and Y workers will only intensify. Also, efforts to retain them will intensify because they are less loyal and more likely to job hop than the baby boomer generation. The average tenure for workers in all age groups has been gradually declining for the past several decades. For workers in Gen X and Y, the average tenure is under three years: 80% of the Gen X and Y workforce have a tenure of five years or less, and fully one-third are in their first year with their employers. Even the promising Gen X and Y employees who enjoy extensive early training and career management often quit before companies can recover their investments in their development.

Gen X and Y workers are also the unhappiest on the job. In a recent survey, this group had the lowest overall satisfaction and workplace engagement levels. They are usually struggling to adjust to the demands of their professional and private lives, more than most employers might imagine. They distrust large organizations (based on their parents' experiences with layoffs), and they often refuse to compromise on work arrangements and workplace style. Employers who expect youthful enthusiasm and a desire to please, so characteristic of the baby boomers, are in for a shock with Gen X and Y's younger workers.

Most employers treat Gen X and Y workers much as their parents had been treated — too often with superficial training, benign neglect, and blind faith. They expect Gen X and Y workers to train diligently, learn the ropes, and wait patiently for advancement opportunities and recognition. Such treatment will dampen the spirit, energy, and ambition of today's younger workers. Inevitably, it will result in a high turnover among the Gen X and Y workers.

Gen X and Y workers want individual responsibility, regardless of their experience. They want to collaborate in decision-making and in their own performance management, they want access to and respect from managers and mentors, and they want to contribute immediately — not gradually, as they earned their stripes. In terms of compensation and benefits, this perspective translates into pay-for-performance, and benefits that are portable. It also translates into cash in hand now, rather than long-term rewards, such as stock options.

The breakdown of the social contract that used to exist between companies and employees coincides with the growing mobility of workers and the portability of their knowledge and worker skills. Many Gen X and Y workers expect no loyalty from their employers nor intend to give it. Many detest the seemingly overwhelming economic and political power of large corporations. This view partly explains why their tenures are brief.

A former manager at a petrochemical company observed that even with the baby boomers, job progression was important. One of their competitors had a greater number of levels in their organization than did his organization. He found his company was continually losing employees to their competitor, simply because individuals had a greater opportunity for promotions. This at least gave them the feeling of career movement. This manager believes that this may be part of the explanation for the Gen X and Y mobility. If they can not find career movement or experience career frustration in their current company, they will move to another company where such opportunity exists.

A few companies offer "stay an extra year" retention bonuses for recent hires, but these rewards can get expensive and often just delay the inevitable. The fundamental challenge goes deeper than retention tactics. Employers must become flexible enough — across work and worker management practices — to meet the changing needs, expectations, and methods of the first truly digital generation in the workforce (especially with Gen Y employees).

However, traditional management methods and styles proved ineffective in meeting these expectations. For example, giving feedback once a year and a formal review just doesn't meet Gen X and Y workers' expectations about the frequency and intensity of communication. Many will be gone or emotionally disengaged from their jobs well before the first year is up. Some of the differences that managers will have to cope with when managing a mixed generation workforce are summarized in Figure 12-1.

In addition to the differences in management styles that will be required, the processes for recruiting and hiring need to adjust to the fact that young workers don't find the conventional employee deal attractive. The emphasis on long-term accumulation of benefits has little appeal. Jobs, training methods, and everyday management practices have to be adjusted to Gen X and Y workers' needs and expectations. Older managers may have difficulty adapting, especially if they don't already understand Gen X and Y workers and don't respect their differences.

Generations and Values

Baby Boomers	Generation X	Generation Y
Competitive	Entrepreneurial	Takes Risks – Breaks Rules
Enjoys Change	Independent and Creative	Enjoys Being Challenged
Hard Work	Quality of Life Focused	Works to Live
Career Progression	Self Managed Careers	Desires a Fun Work Environment
Authoritarian Leadership	Self Directed Leadership	Needs Independence

Figure 12-1

To traditional managers, Gen X and Y workers can appear brazenly demanding and impatient. Not willing to bide their time and earn their stripes, these workers are uncompromising regarding their workplaces and employers. Behind such demands however are genuine differences from the less demanding Baby Boomers. The Gen X / Y worker is:

- Independent, not only intellectually, but also functionally, having grown up fast and managing for themselves from a relatively young age.
- Situational more than structured, and so they feel free to ignore policies and procedures they find restrictive.
- Digital in how they process information and communicate, and sometimes digital at the expense of interpersonal relationships.
- Diverse and comfortable with diversity, so that one size fits all policies and management methods will likely alienate significant numbers of them.

Employers who label these characteristics as having a poor work ethic and expect these workers to outgrow them will suffer endless turnover. Those who respect these differences in thinking, communicating, and problem-solving will try to meet the next generations of workers

Generation "X" and "Y" Retention Actions

- Talk with new hires about how they are "Fitting In"
- Focus on a Fast Start
- Involve Senior Management
- Get them out quickly
- Communicate with them frequently
- Make Retention a real process
  - Involve line managers
- Measure their engagement, mobility, and learning
- Use effective exit interviews
- Evaluate their return
  - Track departed employees
  - Ask and welcome them back

Figure 12-2

at least halfway. They will at least try to understand their preferences, engage their energies, and incorporate their methods.

A lot of today's Gen X and Y workers are also uneasy on the job, not because of any experience or lack of time adjusting to the workday world, but because they seek a different kind of workplace, employment deal, and employer than was typically offered and popularly accepted by the "Baby Boomer" generation. They highly value a congenial, enjoyable workplace, yet — according to Dychtwald, Erickson, and Morison's Workforce Crisis —fewer than 50% have it. They also value the optimal mix of work and life experience, yet they are struggling more — at both work and home — than their managers might assume. They want to believe in their employers, but fundamentally distrust corporations and senior management (again due to their parents experiences).

How can Gen X and Y workers be retained? Figure 12-2 highlights some ideas.

When working with Generation X or Y employees, managers should periodically sit down with them over a cup of coffee and ask if they are comfortable with the work environment. This communication is important because their perception of the workplace will determine whether or not they stay with the company. At this point, at least, any objectionable policies or procedures can be discussed. If a resolution is not available, at least an understanding can be achieved.

These employees should also be focused on a fast start. They do not want to spend a long time in training and would like to be viewed as valuable contributors on a team. The faster they can be brought up to speed and made functional, the more rewarding they will feel their job assignments are.

Senior management should also be involved with Generation X and Y employees to make them feel valued. Frank discussions with these employees may also help senior management understand some of the new workforce paradigms and allow them to proactively implement policies and procedures that will help retain these employees.

Communicating with them about their work — possibly as often as once a week — insures that they don't become disgruntled with their work assignments, believing no one cares. Feedback cannot be as infrequent as yearly reviews, because most Generation X and Y employees will not wait that long to change jobs if they feel undervalued. Good communication will quickly help set expectations for both the employee and the manager; it will help managers understand the employee's perspective on their jobs.

Employee retention should be a priority. A defined process should be in place to insure retention is successful. The HR department should set guidelines (both in timing and topics) for departmental managers to have discussions with their employees. Any time job dissatisfaction is discovered due to company policies and procedures, a further discussion and a mutually satisfactory solution should be achieved.

Companies also want to monitor the involvement and engagement of Generation X and Y employees to ensure they are satisfied with their jobs. If they are engaged in self-directed teams or in projects that encourage them to take the initiatives, retention will be higher. Also, when openings occur in the plant, they should be offered the opportunity to move to another job in the plant, perhaps a more challenging job assignment. They also need to be encouraged to continue to learn, and the opportunities to learn should always be provided to them.

When employees decide to leave, managers should conduct effective exit interviews. These interviews should help the manager understand why employees are leaving. The validity of their reasons should be audited against current company policies to see if there are conflicts, which caused the employee to depart. If there is a particular problem with a company policy, such as one that leads to multiple employees leaving the company, then this information should be captured during the exit interviews. Then recommendations can be made to adjust the policy so that the company won't lose more employees.

Companies should monitor opportunities for good employees to return. If the employees were valued contributors, then they should be tracked. When opportunities arise, they should be asked to come back. If they do return, they should be welcomed back, and not treated as outcasts.

The problems and solutions listed in Figure 12-1 are not all-inclusive. Nevertheless, if the suggestions are followed, the retention rate for Generation X and Y employees should be higher than if they're not followed.

Training Requirements

The manner in which today's Generation X and Y workers learn is radically different from their predecessors. Rather than linear learning from authoritative sources, Generation X and Y workers tend to learn through a process of assembly — that is putting pieces of knowledge and information together from a variety of sources. This means that the training programs for them will need to smaller in unit size, more pointed, and more frequent to cover the same volume of material that was used for the baby boomer generation.

This is where training programs such as one-point lessons, job focused, and equipment specific training will be useful. These types of training programs will be similar to their learning styles and should be more acceptable for their style of learning.

Different economies, industries, and companies will be differently affected by the aging workforce. Projections indicate that 80 percent of the impending global labor shortage will involve skills, and not just the numbers of workers potentially available. Therefore, we will have workers available to do the work, but they will lack the skills to actually perform the work satisfactorily. The competition for skilled labor among countries, regions, industries, and enterprises is already becoming evident, and will significantly increase in the future. Engaging the Generation X and Y employees on the job, while retaining the knowledge and skills of the baby boomer generation, will be critical for companies wanting to survive the next decade.

13

Knowledge Management

We will begin the chapter on knowledge management by looking at the development of specialty skills for critical equipment. Once the development of specialty skills is understood, certification will be discussed. With this level of workforce proficiency achieved, we can consider how to capture this knowledge for use by the future workforce.

Critical Equipment and Specialty Skills

We must first identify critical equipment. We start with equipment that is required for safety or regulatory purposes. Critical equipment is next specified by the importance it has to the production process. As the list of critical equipment grows, another consideration comes into play: does a duty–task–needs analysis exist for the equipment?

Why is such an equipment-specific analysis important? Specialty skills training is usually developed for specific equipment, a serialized spare, or a particular location. Companies may require specialty skills for rebuilding equipment and components or to perform troubleshooting tasks.

In order to develop specialty skills for specific equipment, technicians will need to understand the theory of operation for that equipment. The theory of operation is essentially how the equipment is designed to operate. Any training that focuses on the theory of operation for specific equipment will build on the knowledge and skills the technicians already possess.

Once the technicians understand how the equipment is designed to operate — electrically, mechanically, and hydraulically (with fluid power) — they have a foundation of knowledge that will help them successfully perform maintenance and operational activities on the equipment. Whether they need to repair or rebuilt the equipment, they will know how to perform these activities to design specifications. When technicians develop that level of knowledge about a piece of equipment, they will become more effective maintainers and troubleshooters. Trainers can incorporate much of the theory of operation into flow charts that technicians can use for troubleshooting. All of these components — the technician's existing knowledge, the theory of operation, flow charts for troubleshooting, and so forth — are part of the process of developing specialty skills.

Certification

Managers will want to make sure that their technicians understand enough about how the equipment is supposed to operate so they can be certified in its operation and maintenance. In that way, the managers know that when the technicians are sent out to do a particular job on that specific piece of equipment, the work will be performed correctly. In some cases, employees will need refresher training on special equipment, especially if they have not recently performed work on the equipment (or at all). They may also want to review the theory of operation if the equipment needs to be rebuilt or redesigned. This specialized information, once it is developed, can be reused by any technicians and, even in some cases, operators, if they are to be involved in any maintenance activities on the equipment.

The Disney organization is a classic example of a company that certifies their maintenance personnel for specific maintenance procedures on their ride attractions. They are so disciplined that certain attractions can not operate unless a technician who has been certified on the attraction has signed off that the preventive maintenance tasks have been properly performed on the equipment. It is easy to see why these procedures are important to their business — the consequence of failure is tremendous.

For the same reason, the airline industry is extremely disciplined in its employee certifications. Given the consequence, a failure would be extraordinarily inexcusable. The same consideration exists in the nuclear industry; they can't can't have a meltdown or even begin to think of a nuclear disaster. The public can quickly recall the two disasters that did occur — Three Mile Island and Chernobyl. It's not just that these industry examples have such high visibility that an incident or accident cannot be allowed. The consequences of the accident at Chernobyl are still being felt. Employees in these industries can not just be trained. It is absolutely essential that they be certified to perform equipment specific repairs.

There are still other examples, such as the BP explosion where the individuals involved actually lost their lives. When that event was investigated, multiple errors were discovered. If the employees had been properly trained and had followed proper procedures, the explosion likely would not have happened.

In short, whenever specialty pieces of equipment are being used, the company should require specialty skills. Companies should develop the specialized training correctly so they never have an incident, let alone an accident.

Developing Specialty Skills

The key to developing specialty skills training is applying the duty–task–needs analysis to a specific piece of equipment instead of applying it to a specific skilled trade. For example, instead of asking "What duties does an electrician perform?" start by asking "What duties are performed on this piece of equipment?" The duties can then be divided into craft specific tasks and the training developed according to the guidelines provided earlier in Chapters 5–9.

As noted throughout this text, companies will soon realize that developing the technical skills of the incoming workforce is a priority. Companies must capture the knowledge of their technicians who are here now, and then use that knowledge to develop the training necessary for the next generation of employees.

This book is Volume 5 of the Maintenance Strategy series. If you compare the topical flow that is developed in this series to the development of the employees' technical skills, the importance of building a solid foundation of maintenance practices, such as PMs, MRO stores, and work management processes becomes clear. Good maintenance strategies begin with the basics, and in Volume 1, the Preventive Maintenance program was discussed in detail.

In any PM program, about 50% of all equipment failures begin with the neglect of the basics, such as proper lubrication practices, proper fastening procedures, and good visual inspections. Companies want to insure that their technicians have the appropriate technical knowledge, so that these fundamentals are in place and that the technicians are consistent in how they apply the maintenance basics. When maintenance basics are not applied correctly, companies need to identify the skill deficiencies. Therefore, evaluating the technicians' skills is critical. Once the company understands their current level of skills, technicians and other trainees can be enrolled in the proper training programs. These programs will enable the technicians not only to do their jobs correctly, but also to work safely themselves and with their coworkers so no one has an incident or accident. In addition, these programs will allow them to maintain the equipment competently, with zero rework.

Plant and facility equipment are continually being upgraded or replaced. In turn, technicians will continually need their training upgraded. Any time equipment or processes are introduced or altered companies should establish a defined process to capture the change; thus, if any new skills are required or if procedure changes, the training program is appropriately updated. Also, companies need to provide periodic refresher training so that all individual technicians remain current on those skills they do not often apply.

The training program and materials also need to be upgraded continually, whether they are for apprentice training, specialty skills development, or skills refreshment. This need for upgrading gives trainers additional tasks because new materials are being written every day — equipment manuals from vendors, new editions of textbooks, and interactive computer-based training programs. Companies want their trainers to review these new materials continually. The best training materials will then be in place so that the technicians and operators have the greatest opportunity to improve their skills.

Knowledge Capture

As was highlighted in Chapter 1, many of the more skilled and knowledgeable technicians will be eligible to retire from the workforce in the next few years. How will companies prevent the loss of knowledge and skills when this occurs?

In order to understand knowledge capture, it is important to understand three terms that are common in an industrial setting: data, information, and knowledge.

Data are a set of particular and objective facts about an event or even be a structured record of a transaction. Data do not say anything meaningful; they are just raw facts and figures. For data to be useful, they must be converted into information.

Peter Drucker defines information as "data endowed with relevance and purpose." Information comes from the root inform, which means that it is something that shapes or changes the person who receives it. The person who receives the data decides if it is information or just "data noise." What differentiates data from information is a subjective judgment. Information helps managers:

• Run a business more efficiently and effectively
• Make more effective decisions
• Make changes in business direction

One of the biggest problems in differentiating information from noise is that there is often too much information, resulting in overload. In these cases, it is virtually impossible to make sense of the information. It loses its relevance and purpose, and therefore becomes almost useless, converting back into simply data. The need to manage information is one of the reasons companies struggle with performance indicators — they are trying to convert data into information, and then focus that information.

Knowledge is actionable information. This means we need relevant information in the right place, at the right time, in the right context, and in the right way so it can be used to make business decisions. Knowledge is a key resource in intelligent decision making, forecasting, design, planning, diagnosis, analysis, evaluation, and intuitive judgment making. Knowledge involves information blended with experience, successes, failures, and learning over time. Thus knowledge is information that is stored or captured in its context. Knowledge allows for making predictions, casual associations, or predictive decisions about what to do — unlike information which simply provides facts.

Data	Information	Knowledge

Building control system captures the fact that the building temperature is 80 degrees	The computer monitoring system recognizes that this data is out of the 74 to 76 degree specification	This information is acted on by the technicians who has technical skills to make the adjustment

Figure 13-1

The relationship among data, information, and knowledge is highlighted in Figure 13-1. In this figure, a building sensor indicates that the temperature has reached 80 degrees. This is a piece of data. Other non-relevant data are the humidity is 40%, the barometric pressure is 30.00 in inches of mercury, etc. The data becomes information, once it is given parameters against which it is measured. The building control system has been set to monitor set points of 74–76 degree specification. Now, it can be recognized that the data is not within a given range. Armed with this relevant information, the technician needs to take action. The technicians, who are aware of the temperature being out of specification, combine their experience and the information to come to a predictive decision and make the correction.

This process leads to the understanding of explicit and tacit knowledge. Explicit knowledge can be captured in electronic or hardcopy databases. It typically includes areas such as "if this, then that" or "cause and effect" knowledge. Tacit knowledge, by contrast, is not easily captured. Tacit knowledge involves understanding organizational social and cultural paradigms. In other words, tacit knowledge involves not only what to do, but how to get it done in the current organizational setting. In the case of the building temperature in Figure 13-1, the explicit knowledge is the technician's knowledge of how to adjust the temperature. The tacit knowledge is being aware of who in the organization they need to contact to gain permission to make the adjustment.

Examples of Explicit Knowledge

As highlighted in Volume 4 of the Maintenance Strategy Series, the Computerized Maintenance Management System (CMMS) or Enterprise Asset Management (EAM) system is a repository of an incredible amount of maintenance, reliability, and equipment data. In addition, there are other systems that companies may rely on, such as Building Automation Control (BAC), Manufacturing Execution Systems (MES), and Distributed Control Systems (DCS) and others. The data these systems collect and store can be accessed by reporting tools and used to develop "if this, then that" scenarios and decision trees. This process begins the transformation of the data (contained in the various systems) into information.

For example, it is possible to plot the pressure on both sides of a filter. The listing of pressure points is data; by itself the listing doesn't have a purpose. But next we can compare these points to earlier ones to measure the pressure drop across the filter. Now the data have relevance and become information about the change in pressure. It can be determined when to change the filter. In turn, Subject Matter Experts (SMEs), drawing on their experience, determine that when there is a pressure drop of more than 5 PSI (Pounds per Square Inch), the filter should be replaced. The data (the pressures) and the information (changes in pressure) have been transformed into actionable information or explicit knowledge (when the pressure drops > 5 PSI).

This example is a rather simple one. However, there are virtually hundreds of these types of data–information–knowledge transformations that take place in any plant or facility. The problem is finding the time to sit down with the appropriate SMEs to collect the information necessary that completes the transformation from data into knowledge. This type of information is what must be collected from the older SMEs in companies today before they retire. If this knowledge is not collected, recorded, and used in a decision-making tool, then it will be lost forever. The cost to have new employees learn and then apply this information on their own will be prohibitive. The cost of re-learning this information may be the cost that finally decides if a company stays competitive or goes out of business.

It is not that difficult to collect the knowledge from the SMEs while they are currently on the job. The data can be collected in the CMMS or EAM system that is used to generate their work orders. Most systems have codes for every work order closure. These codes are generally called P-C-A codes — they are used to say what the Problem was, what Caused the problem, and what Action was taken to correct the problem.

Once this data is captured, we can easily write reports such as the top ten equipment problems that are occurring in the plant. Then using this information, identify the most common causes of the problems, and determine the action that was taken to correct the problem. Over time, this information can be used to build a database for decision making. For example, I may have hydraulic pumps that cavitate. When considering the likely causes, they may be:

- A clogged inlet filter
- An obstruction in the inlet
- Overheated hydraulic fluid

Over time, when looking at pump cavitation, I may find the following data:

- A clogged inlet filter — 20 occurrences
- An obstruction in the inlet — 30 occurrences
- Overheated hydraulic fluid — 50 occurrences

If the database is properly constructed, it would be possible to ask for a probability calculation, which would report that there is:
- A 50% probability that there is overheated hydraulic fluid
- A 30% probability that there is an obstruction in the inlet
- A 20% probability that there is a clogged inlet filter

This process allows the data from the CMMS/ EAM system to be transformed into actionable information or explicit knowledge.

The next step is to work with the SMEs to devise a way to permanently eliminate the identified problem. Our goal is to build a process for capturing the SMEs' knowledge so that we can increase the plant's capacity and lower the plant's overall cost to produce.

Artificial Intelligence and Explicit Knowledge

Artificial intelligence systems are commercially available today. If companies were efficient at collecting the proper data and combining it with the information from the SMEs, they could truly capture the explicit knowledge that exists in their organizations today. The problem with today's artificial intelligence (AI) systems has been loading the rules engines — that is, combining the hard data from the internal company systems with the experience and explicit knowledge of the SMEs. Most companies are not dedicating the resources necessary to begin and complete this type of project. Not dedicating these resources is one of the most catastrophic mistakes that a company can make today.

Although artificial intelligence may seem like an idea out of a science fiction movie, this type of initiative has been undertaken since the 1980s. A major company developed an artificial intelligence program for troubleshooting problems on their production press equipment for the automotive industry. The AI system was called "Project Charlie." I personally saw the program operated and it was very impressive. It began with an individual named Charlie, who was the best press troubleshooter in the entire company. In fact, the company was flying Charlie around the country to work on presses that were causing problems.

Then some forward-thinking individual realized that Charlie was 59 years old and would soon retire. The company decided to do something before Charlie retired and they lost all of his expertise. So they sent Charlie and his wife to Southern California to a think tank for two years. There, AI experts gathered as much knowledge from Charlie as they could and put it into a computer. The result was an interesting AI tool that was actually demonstrated at several different advanced manufacturing shows in the late 1980s early 1990s. It was called "Project Charlie" and it worked like this:

1. You would be asked to give some data input as to what the press was doing incorrectly.

2. The program would come back with a direction "Based on the data you provided, go out and gather this data." This request was typically pressure readings, dimensional tolerances, operational settings, etc.

3. You returned and input this data in the system.

4. The system would then continue with the next direction, "Based on the data you provided, go out and gather this additional data." or The system sometimes asked instead for some data to be verified, if the program determined the readings were suspect.

5. After entering the second (and sometimes third) set of data, the program would provide feedback such as "There is a 50% probability of this problem; a 20% probability of this problem; and a 15% probability of this problem" and then give the results in descending order.

The AI system tracked each press it was used to diagnose. When the system was used again in a plant, it would ask for the equipment identifier of the equipment. You enter that information and what problem was found; therefore, the system would learn each specific piece of equipment. This program was eventually set aside when the people who ran it retired.

Anyone who believes that this type of program will never gain acceptance should realize that AI systems are the future for knowledge management in the technical trades. The concept may sound a little bit like Star Trek at first, but that's indeed where the industry is headed. The military has already undertaken extensive work in this area. If this area is of interest to you, look for a book titled Artificial Intelligence in Maintenance (Richardson, Noyes Publications.) The book is based on what the military has already done in AI systems.

The American Society for Training and Development (ASTD) has published much information on knowledge management. The ASTD focuses on five categories of knowledge management activities. These five categories are:

- Defining intellectual capital
- Creating intellectual capital
- Capturing intellectual capital
- Sharing intellectual capital
- Using intellectual capital

Although it is beyond the scope of this text to cover all five of these categories, it is interesting to examine capturing intellectual capital.

From ASTD's perspective, one of the first steps in capturing intellectual capital is building best practice databases. Another way to capture intellectual capital is by creating knowledge repositories. A third way is by compiling process documentation and reengineering. A fourth way is by writing manuals.

Reflecting on the topics covered in this text, it is easy to see that the steps to develop a technical training program can also be used to collect or capture intellectual capital. Each company must evaluate their technical training programs and consider how they can be further utilized to prepare for the coming generational and educational changes in the workforce. The three elements of "the perfect storm" described in Chapter 1 are coming together. Companies throughout the world will have very serious problems with technical skills in maintenance, which may lead to a noncompetitive position in their marketplace.

I hope this text will not only create an awareness of the problem, but also provide ideas that lead to solutions within your organization.

Appendix A

Mechanical Task Descriptions

01.00 **Demonstrate Basic Employment skills and habits**
01.01   Identify employment opportunities.
01.02   Apply employment-seeking skills.
01.03   Interpret employment capabilities.
01.04   Demonstrate appropriate work behavior.
01.05   Maintain safe and healthy environment.
01.06   Maintain a business-like image.
01.07   Maintain working relationship with others.
01.08   Communicate on the job.
01.09   Adapt to change.
01.10   Demonstrate knowledge of business.
01.11   Perform mathematical skills.
01.12   Demonstrate and maintain ethical behavior.
01.13   Compile a portfolio.

02.00 **Demonstrate knowledge of general safety orders**
02.01   Explain the purpose(s) of the OSHA Act.
02.02   Apply shop safety rules and procedures.
02.03   Apply personal safety rules and procedures.
02.04   Apply electrical safety rules and procedures.
02.05   Apply fire safety rules and procedures.
02.06   Report injuries in a timely manner.
02.07   Recognize and report unsafe acts.
02.08   Recognize and report unsafe conditions.
02.09   Demonstrate proper use of personal protective equipment (PPE).
02.10   Demonstrate proper lifting methods.
02.11   Apply hazardous wastes rules and procedures.
02.12   Demonstrate knowledge of emergency evacuation routes, master switch and lockout location.

03.00 **Demonstrate ability to read engineering drawings.**
03.01   Explain the purpose of engineering drawings.
03.02   Explain and interpret machine parts and machine drawings.

205

03.03 Read machine assembly drawings.

03.04 Develop sketches.

03.05 Compute materials from engineering drawings.

03.06 Interpret building drawings.

03.07 Read and interpret drawing schematics and symbols.

03.08 Identify common features and differences of schematics.

03.09 Identify electrical wires and connections.

03.10 Read electrical diagrams.

03.11 Identify piping systems, projection, joints, valves, and symbols.

03.12 Read a piping schematic.

03.13 Identify fluid power system component symbols and interpret diagrams.

03.14 Interpret air conditioning and refrigeration system and subsystem schematics.

03.15 Identify symbols for welds.

04.00 Demonstrate Shop Skills

04.01 Add, subtract, multiply, and divide positive and negative numbers.

04.02 Add, subtract, multiply, and divide fractions.

04.03 Change mixed numbers to decimals.

04.04 Compare numbers and calculate ratios.

04.05 Demonstrate understanding of geometric functions.

04.06 Solve algebraic equations.

04.07 Explain the properties of triangles.

04.08 Apply trigonometric functions to layout and installation situations.

04.09 Demonstrate understanding of metric and linear measurement.

04.10 Measure bulk materials.

04.11 Use plumbing codes to determine materials.

04.12 Explain techniques of measuring motion, forces, fluids, electricity, and temperature.

04.13 Explain the mechanical and chemical properties of ferrous and non-ferrous metals.

04.14 Explain industrial manufacturing processes.

04.15 Explain the industrial use of non-metallic solids, liquids, and gases.

04.16 Explain the precautions required when using toxic or flammable materials.

04.17 Demonstrate use and maintenance of personal protective equipment.

05.00 Demonstrate Using and Maintaining Hand Tools

05.01 Demonstrate the use of rules, tapes, calipers, and micrometers.

05.02 Demonstrate the use of wrenches and screwdrivers.

05.03   Demonstrate the use and care of pipefitting tools.

05.04   Demonstrate the use of line clearing equipment.

05.05   Demonstrate the use of equipment used to bend and assemble rigid electrical conduit.

05.06   Demonstrate the use and care of woodworking tools, including saws, planes, drills, hammers

05.07   Demonstrate the use and care of sheet metal tools, including sheet metal gauges, layout

05.08   Demonstrate proper metalworking bench skills, including the use of vices, hacksaws, files

05.09   Demonstrate the use and care of ropes, slings, pulleys, and block and tackle.

05.10   Demonstrate the use and care of test and safety equipment.

06.00   Demonstrate Using and Maintaining Portable Power Tools

06.01   Demonstrate the use and maintenance of light-duty and heavy-duty drills.

06.02   Demonstrate the use and maintenance of electric hammers.

06.03   Demonstrate the use and maintenance of pneumatic drills and hammers.

06.04   Demonstrate the use and maintenance of power screwdrivers and impact wrenches.

06.05   Demonstrate the use and maintenance of linear motion saws.

06.06   Demonstrate the use and maintenance of circular saws.

06.07   Demonstrate the use and maintenance of routers and planes.

06.08   Demonstrate the use and maintenance of belt, pad and disc sanders.

06.09   Demonstrate the use and maintenance of grinders and shears.

06.10   Demonstrate the use and maintenance of explosive actuated tools.

06.11   Sharpen tools using a bench grinder.

07.00   Demonstrate Using Stationary Shop Equipment

07.01   Demonstrate the use of mechanical presses.

07.02   Demonstrate the use of hydraulic presses.

07.03   Demonstrate the use of drill presses.

07.04   Demonstrate the use of bench grinders.

07.05   Demonstrate the use of power hack saws, cut-off saws or chop saws.

07.06   Demonstrate the use of band saws.

07.07   Demonstrate the use of pipe threaders.

07.08   Demonstrate the use of power metal brakes.

07.09   Demonstrate the use of power shears.

08.00   Demonstrate Basic Troubleshooting Skills

08.01   Explain the importance of maintenance.

08.02   Explain and demonstrate troubleshooting procedures.

08.03   Identify aids to troubleshooting.

08.04   Demonstrate knowledge of the safety rules for troubleshooting and repair procedures.

08.05   Maintain troubleshooting and repair records.

08.06   Use manufacturer's manuals, schematics, and troubleshooting charts, as well as general

09.00   Demonstrate Using Gas Welding / Cutting Equipment

09.01   Identify gas welding and cutting equipment and accessories.

09.02   Identify personal protective equipment required for welding and cutting.

09.03   Explain capillary attraction as it applies to metal joining.

09.04   Demonstrate proper lighting, adjusting, and shutting down of a gas torch.

09.05   Layout and cut mild steel.

09.06   Braze mild steel.

09.07   Braze cast iron.

09.08   Solder non-ferrous metals.

10.00   Demonstrate Using Arc Welding / Cutting Equipment

10.01   Set up and adjust a shielded metal arc welder.

10.02   Identify and select electrodes.

10.03   Strike, maintain and restart an S.M.A.W. arc.

10.04   Weld straight bead in flat position.

10.05   Weld weave bead patterns.

10.06   Weld build-up pads.

10.07   Weld basic joints in flat position (1G and 1F).

10.08   Weld basic joints in horizontal position (2G and 2F).

10.09   Weld basic joints in vertical position (3G and 3F).

10.10   Weld basic joints in overhead position (4G and 4F).

10.11   Weld cast iron.

10.12   Weld alloy steels.

10.13   Build up shaft or round surface.

10.14   Weld aluminum.

10.15   Hard surface metals with S.M.A.W.

10.16   Set up a gas tungsten arc welder.

10.17   Select and prepare a tungsten electrode.

10.18   Strike and maintain a G.T.A.W. arc.

10.19   Weld mild steel in all positions (1F thru 4F).

10.20   Weld stainless steel in all positions (1G thru 4G).

10.21   Weld aluminum in all positions (1G thru 4G).

10.22   Set up and adjust G.M.A.W. and F.C.A.W.

10.23   Weld weave bead patterns using G.M.A.W. and F.C.A.W.

10.24   Weld basic joints in flat, horizontal, and vertical positions.

10.25   Set up air carbon arc cutting equipment.

10.26   Gouge, cut and pierce metals using air carbon arc.

11.00   Demonstrate Knowledge of Basic Electricity & Electronics

11.01   Define common terms used in electricity and electronics.

11.02   Discuss the National Electrical Code.

11.03   Explain the nature of static electricity.

11.04   Explain methods used to measure and control static electricity.

11.05   Explain the theory of magnetism.

11.06   Describe the industrial uses of magnets and electromagnets.

11.07   Explain the purpose and use of transformers.

11.08   Explain Ohm's Law.

11.09   Use instruments which measure current, resistance, and potential difference.

11.10   Explain the fundamentals of DC circuits.

11.11   Explain the use of DC circuits in motors and generators.

11.12   Explain the use and function of electrical and electronic control equipment.

11.13   Discuss programmable controllers.

11.14   Explain the differences between AC and DC circuits.

11.15   Demonstrate knowledge of the instruments used to measure electrical circuits.

11.16   Measure load in three phase circuits.

11.17   Install electric motors.

11.18   Demonstrate knowledge of troubleshooting procedures for electric circuits and control systems.

11.19   Troubleshoot DC motors.

11.20   Troubleshoot AC motors.

11.21   Troubleshoot lighting systems.

12.00   Demonstrate Knowledge of Basic Mechanical Theory

12.01   Demonstrate an understanding of measuring systems and ratios.

12.02   Explain working forces of torque, tension, and compression.

12.03   Explain the laws of motion.

12.04   Explain how to calculate work.

12.05   Explain the function of simple machines including levers, inclined plane, wedge wheel.

12.06   Explain the types of power and the method of producing power including compound gears.

12.07   Calculate volume mathematically and by displacement.

12.08   Explain the laws of friction.

12.09   Explain horsepower and how to calculate

13.00   **Demonstrate Knowledge of and Usage of Lubricants**

13.01   Explain the function of lubricants (both oil and grease).

13.02   Explain the properties of oil lubricants and factors determining the selection of lubricants

13.03   Identify the types and functions of lubricant additives.

13.04   Describe the types of circulating oils and their purposes.

13.05   Describe lubricating systems, including the charts and methods used

13.06   Demonstrate proper grease application.

13.07   Demonstrate proper lubricant storage and handling.

13.08   Lubricate a piece of industrial equipment.

13.09   Maintain lubricating automated systems

13.1   Explain the various properties associated with lubricants. (i.e. viscosity, melting point, etc.)

14.00   **Demonstrate Knowledge of and Maintenance of Drive Components**

14.01   Install a solid coupling.

14.02   Install a jaw coupling.

14.03   Install a molded rubber coupling.

14.04   Install a chain type coupling.

14.05   Identify and install a clutch.

14.06   Install V-belts and adjust tension.

14.07   Install and adjust a V-belt and manually adjustable sheaves.

14.08   Install a V-belt with spring loaded adjustable sheaves.

14.09   Describe and explain the purposes and advantages of a chain drive systems.

14.10   Explain the function of speed reducers.

14.11   Explain the function of gears and variable speed reducers.

14.12   Install and align shafts.

14.13   Mount drive sprockets and chains.

14.14   Mount sheaves and pulleys.

14.15   Mount and align gears on open gear drives.

14.16   Install a mechanical clutch system.

14.17   Install adjustable speed drives.

14.18   Troubleshoot adjustable speed drives.

14.19   Explain the operation of fluid couplings.

14.20   Install fluid couplings.

14.21   Install torque converters.

14.22   Perform preventive maintenance on drive components.

14.23   Calculate gear ratios, gear speeds

14.24   Install and maintain gear boxes.

14.25   Install chain drives

14.26   Install and adjust chain drives gears and chain

14.27   Demonstrate ability to cut chain to proper length

14.28   Install and maintain clutches.

14.29   Install and maintain brakes

15.00   Demonstrate Knowledge of and Maintenance of Bearings

15.01   Identify common bearing types and their advantages.

15.02   Mount, square, and align anti-friction bearings.

15.03   Identify specialized bearings, their applications and characteristics, including: thrust, tandem, etc.

15.04   Identify and select bearing seals for specified applications.

15.05   List rules for good bearing lubrication.

15.06   Explain bearing load, wear patterns, & maintenance.

15.07   Explain the use of cross-reference manuals in bearing maintenance and repair

15.08   Explain the use of bearing forecast maintenance systems.

15.09   Remove, inspect, and replace a plain journal bearing.

16.00   Demonstrate Knowledge of and Maintenance of Seals and Pumps

16.01   Determine pump capacity and system requirements.

16.02   Identify packing and seal requirements.

16.03   Explain the operating principles of centrifugal, propeller and turbine rotary, reciprocating pumps

16.04   Disassemble and reassemble a pump.

16.05   Perform pump maintenance.

17.00   Demonstrate Knowledge of and Maintenance of Piping Systems

17.01   Identify the components of a piping system.

17.02   Explain the maintenance features of both metallic and non-metallic piping systems.

17.03   Explain valve operation and maintenance.

17.04   Explain the use and maintenance of strainers, filters, and traps in piping systems.

17.05   Bend and join copper tubing.

17.06   Bend and join steel tubing.

17.07   Join common fittings.

17.08   Join metallic pipe.

17.09   Join plastic pipe.

18.00   Demonstrate Knowledge of and Maintenance of Hydraulic Systems

18.01   Explain Pascal's Law.

18.02   Explain Bernoulli's Principle.

18.03   Explain how heat and pressure relate to power and transmission.

18.04   Describe the physical and chemical properties of a fluid.

18.05   Install and maintain a contaminant removal system.

18.06   Explain the operation and use of heat exchangers.

18.07   Determine reservoir requirements.

18.08   Classify and select pumps for specific applications.

18.09   Compute hose requirements.

18.10   Install hydraulic lines.

18.11   Install and maintain control valves and servo-type valves.

18.12   Install and maintain linear actuators

18.13   Install and maintain rotary actuators

19.00   Demonstrate Knowledge of Troubleshooting Hydraulic Systems

19.01   Read a hydraulic schematic.

19.02   Connect electrically controlled valves.

19.03   Explain hydraulic system troubleshooting techniques.

19.04   Repair and replace hydraulic valves.

19.05   Repair and replace hydraulic cylinders.

19.06   Repair and replace hydraulic pumps and motors.

19.07   Cut, flare, and bend hydraulic tubing.

19.08   Install hydraulic components.

20.00   Demonstrate Knowledge of and Maintenance of Air Compressors

20.01   Explain the relationship of force, weight, mass, pressure, and density in a pneumatic system.

20.02   Explain the operation of reciprocating compressors.

20.03   Explain the operation of rotary compressors.

20.04   Explain primary and secondary air treatment.

20.05   Explain the operation of compressor valves, cylinders, and motors.

20.06   Explain the difference between absolute and gauge pressures.

21.00   Demonstrate Knowledge of and Maintenance of Pneumatic Systems

21.01   Identify the schematic symbols and diagrams used in pneumatic systems.

21.02   Diagram an air supply system.

21.03   Install pneumatic system components.

21.04   Explain pneumatic system maintenance techniques.

21.05   Explain pneumatic system troubleshooting procedures.

21.06   Troubleshoot air compressors.

21.07   Troubleshoot pneumatic control valves.

21.08   Troubleshoot, repair, and install control valves.

21.09   Troubleshoot air motors.

21.10   Troubleshoot air dryers.

21.11   Install and maintain control valves and servo-type valves.

21.12   Install and maintain linear actuators

21.13   Install and maintain rotary actuators

21.14   Troubleshoot air reservoirs.

21.15   Install and maintain flow control valves

22.00   Demonstrate Knowledge of Pollution Control Systems

22.01   Explain the operation of air pollution control systems.

22.02   Explain the operation of water pollution control systems.

22.03   Explain the operation of solid waste pollution control systems.

22.04   Explain the operation of noise pollution control systems.

22.05   Explain the basic philosophy of "right to know" legislation.

22.06   Explain how to properly report on hazardous spills and releases.

23.00   Demonstrate Knowledge of and Performance Rigging Functions

23.01   Estimate the weight of a load.

23.02   Find the center of gravity.

23.03   Explain procedures for moving and installing new equipment.

23.03   Identify the rigging and slings used in maintenance work.

23.04   Explain safety inspection procedures for rigging, ropes, and slings.

23.05   Explain how to determine if rigging equipment inspection is current.

23.06   Identify rope fiber types.

23.07   Tie rigging knots, bends, and hitches.

23.08   Identify types of wire rope.

23.09   Cut and seize wire rope.

23.10   Install wire rope eyes, sockets, and hooks.

23.11   Identify cranes and hoists.

23.12   Splice wire rope.

23.13   Erect a scaffold and install planking.

23.14   Raise a ladder.

23.15   Rig lifebelts and life nets.

24.00   Demonstrate Knowledge of and Perform Installation of Equipment

24.01   Explain relocation procedures for new equipment in an existing facility.

24.02   Explain the use of anchors and isolators.

24.04   Explain leveling and aligning procedures.

24.05   Explain test run guidelines.

24.06   Explain safety precautions for equipment installation procedures.

24.07   Explain grouting procedures.

25.00   Demonstrate Knowledge of and Perform Machine Shop Turning Operations

25.01   Identify the principal parts of an engine lathe.

25.02   Demonstrate the safe and proper use of lathes and attachments.

25.03   Perform turning operations.

25.04   Perform boring operations.

25.05   Perform drilling and reaming operations.

25.06   Perform internal and external threading operations.

26.00   Demonstrate Knowledge of and Perform Machine Shop Milling Operations

26.01   Identify types of milling machines and tooling.

26.02   Determine spindle speed, feed rates, and direction of feed.

26.03   Perform external milling operations.

26.04   Perform angular milling operations.

26.05   Perform internal milling operations.

26.06   Slab mill a work piece.

26.07   Slot on a horizontal milling machine.

26.08   Mill a keyway.

27.00   Demonstrate Knowledge of Machine Shop Jobs

27.01   Determine sequence of work on a specified project.

27.02   Determine tolerances and finishes.

27.03   Explain the variables that affect job efficiency.

27.04   Explain the use of the Machinery Handbook.

28.00   Demonstrate Knowledge of Computerized Maintenance Management Systems

28.01   Demonstrate knowledge of manual record keeping practices.

28.02   Demonstrate knowledge of electronic record keeping practices.

28.03   Complete a work order.

28.04   Complete an internal requisition.

28.05   Complete an external requisition.

28.06   Use an electronic drawing storage system.

28.07   Define and explain scheduled maintenance.

28.08   Define and explain planned maintenance.

28.09   Define and explain breakdown maintenance.

28.10   Explain the reasons for keeping maintenance records.

28.11   Explain the reasons for keeping cost records.

28.12   Demonstrate basic computer literacy.

28.13   Define statistical process control (SPC).

APPENDIX B

SUGGESTED READING LIST

ASTD. The ASTD Handbook for Technical and Skills Training. ASTD: Alexandria, VA, 1985, 1986. http://store.astd.org/Default.aspx?tabid=141

Carnevale, Anthony P., Leila J. Gainer, and Eric R. Schulz. Training the Technical Work Force (The ASTD Best Practices Series). Jossey Bass, 1990. http://store.astd.org/Default.aspx?tabid=141

Cheney, Scott. Excellence in Practice, Volume 2. ASTD: Alexandria, VA 1998.

Dennis, Pascal. Getting the Right Things Done. Lei Press: Cambridge MA, 2006.

Dychtwald, Ken, et al. The Workforce Crisis. Harvard Business School Press, 2006.

Gandossy, Robert et al. Workforce Wake-up Call. John Wiley and Sons, 2006.

Gordon, Edward E. The 2010 Meltdown: Solving the Impending Jobs Crisis. Praeger, 2005.

Huselid, Mark A., Brian E. Becker, and Richard W. Beatty. The Workforce Scorecard. Harvard Business School Press, 2005.

Leibold, Marius and Sven Voelpel. Managing the Aging Workforce. John Wiley and Sons, 2006.

Mager, Robert F. The New Mager Six Pack. CEP Press, 1992.

Mager, Robert F. What Every Manager Should Know About Training. CEP Press, 1999.

Markova, Dawna. Open Mind. Conari Press, 1996.

Phillips, Jack J. Return on Investment in Training and Performance Improvement Programs. Butterworth-Heinemann, 2003.

Phillips, Jack J. and Patricia Pulliam Phillips. ROI at Work. ASTD: Alexandria, VA, 2005. http://store.astd.org/Default.aspx?tabid=141

Sharpe, Cat (editor). The Complete Guide to Technical and Skills Training. ASTD: Alexandria, VA, 1998. http://store.astd.org/Default.aspx?tabid=141

INDEX